W9-ABN-166

Solution of boundary value problems by the method of integral operators

D L Colton

University of Strathclyde

Solution of boundary value problems by the method of integral operators

Pitman Publishing

LONDON · SAN FRANCISCO · MELBOURNE

PITMAN PUBLISHING LIMITED
39 Parker Street, London WC2B 5PB

FEARON-PITMAN INC.
6 Davis Drive, Belmont, California 94002, USA

Associated Companies
Copp Clark Ltd, Toronto
Pitman Publishing Co. SA (Pty) Ltd, Johannesburg
Pitman Publishing New Zealand Ltd, Wellington
Pitman Publishing Pty Ltd, Melbourne

First published 1976
Reprinted 1977

AMS Subject Classifications: (main) 35A20, 35A35, 35A40
(subsidiary) 35C15, 35J25, 35K20, 35R30

Library of Congress Cataloging in Publication Data

Colton, David L. 1943–
Solution of boundary value problems by the method of integral operators.

(Research notes in mathematics; 6)
"The present set of lectures was given during the academic year 1974–75 while the author was a guest professor at the University of Konstanz."
Bibliography: p. 144
1. Boundary value problems—Numerical solutions.
2. Initial value problems—Numerical solutions.
3. Differential equations, Partial—Numerical solutions.
4. Integral operators. I. Title. II. Series.
QA379.C64 515′.35 76–15190

© D. L. Colton 1976

All rights reserved. No part of this publication may be reproduced, stored in a retrieval system, or transmitted in any form or by any means, electronic, mechanical, photocopying, recording and/or otherwise without the prior written permission of the publishers. The paperback edition of this book may not be lent, resold, hired out or otherwise disposed of by way of trade in any form of binding or cover other than that in which it is published, without the prior consent of the publishers.

Reproduced and printed by photolithography
in Great Britain by Biddles Ltd, Guildford, Surrey

ISBN 0 273 00307 0

1591326

Preface

These lecture notes are intended to be a companion volume to [8]. In [8] a general survey was given of the analytic theory of partial differential equations, with particular emphasis on improperly posed initial value problems and the analytic continuation of solutions to partial differential equations. The use of integral operators to solve boundary value and initial-boundary value problems arising in mathematical physics was discussed only briefly. In the present set of notes this topic now becomes the main theme, and the interplay between analytic continuation and the approximation of solutions to partial differential equations is developed in some detail. With the idea that these two sets of lectures should be read together, we have minimized overlapping topics, while at the same time keeping each set of lectures self-contained. Indeed the only topics common to [8] and the present volume are integral operators for elliptic equations in two independent variables (which is treated in considerably more detail in the present set of notes) and the inverse Stefan problem for the heat equation in one space variable (which occupy only a few pages in both [8] and the present volume).

The present set of lectures was given during the academic year 1974-75 while the author was a Guest Professor at the University of Konstanz. The prerequisites for the course were a one semester course in partial differential equations and a one semester course in analytic function theory. I would like to particularly thank Professor Wolfgang Watzlawek and the Fachbereich Mathematik of the University of Konstanz for their hospitality and financial support during the year, and also the Air Force Office of Scientific Research for their continued support of my research efforts through AFOSR Grant 74-2592. Appreciation is also given to Mrs. Norma Sim of the University of Strathclyde for her patient and careful typing of the manuscript.

Contents

Introduction

The simplest example of the type of problem we will want to consider in these lectures is the following approach for approximating solutions of the Dirichlet problem for Laplace's equation. Let D be a bounded simply connected domain in $\mathbb{R}^2$ with analytic boundary ∂D. We wish to approximate (in the maximum norm) the solution of ($u \in C^2(D) \cap C^o(\bar{D})$)

$$u_{xx} + u_{yy} = 0 \qquad \text{for } (x,y) \in D \tag{1}$$

$$u(t) = f(t) \qquad \text{for } t \in \partial D \tag{2}$$

where $f(t) \in C^o(\partial D)$. From the maximum principle it suffices to approximate the solution of (1), (2) for f(t) analytic. In this case u(x,y) is in fact a solution of (1) in a domain $\tilde{D} \supset \bar{D}$($\bar{D} = D \cup \partial D$).

We have

$$u(x,y) = \text{Re}\{\phi(z)\} \tag{3}$$

where $\phi(z)$ is an analytic function of $z = x+iy$ in $\tilde{D}$ and hence by Runge's theorem the set

$$\begin{aligned} u_{2n}(x,y) &= \text{Re}\{z^n\} \\ u_{2n+1}(x,y) &= \text{Im}\{z^n\} \end{aligned} \tag{4}$$

is a complete family of solutions to (1) in $\tilde{D}$, i.e. for every compact subset $B \subset \tilde{D}$(in particular for $B=\bar{D}$) and $\varepsilon>0$ there exist constants $a_1, \ldots, a_N$ such that for N sufficiently large

$$\max_B \left| u - \sum_{n=0}^{N} a_n u_n \right| < \varepsilon \ . \tag{5}$$

Now orthonormalize the set $\{u_n\}$ in the L^2 norm over ∂D to obtain the complete set $\{\phi_n\}$, i.e.

$$\int_{\partial D} \phi_n \phi_m = 0 \qquad \text{for } n \neq m \tag{6}$$

$$\int_{\partial D} |\phi|^2 = 1.$$

Let

$$c_n = \int_{\partial D} f\phi_n \; . \tag{7}$$

Let D_o be a compact subset of D. From the representation of the solution of (1), (2) in terms of the Green's function it is seen that if

$$\int_{\partial D} |f - \sum_{n=0}^{N} c_n \phi_n|^2 < \varepsilon \tag{8}$$

then

$$\max_{D_o} |u - \sum_{n=0}^{N} c_n \phi_n| < M\varepsilon \tag{9}$$

where $M = M(D_o)$ is a constant. Hence an approximate solution to (1), (2) on compact subsets of D is given by

$$u^N = \sum_{n=0}^{N} c_n \phi_n \quad . \tag{10}$$

Since each ϕ_n is a solution of (1), error estimates can be found by finding the maximum of $|f-u^N|$ on ∂D and applying the maximum principle.

Remark: The assumption that ∂D is analytic is made to avoid technical approximation arguments. However the above method remains valid under much weaker assumptions, e.g. ∂D is Hölder continuously differentiable. We will discuss this in more detail during the course of these lectures.

We have now reduced the problem of constructing an approximate solution to (

(2) to a simple one of quadrature, a method which in fact is well suited to use on a digital computer.

The representation (3) can also be used to obtain a method for approximating the solution to the Dirichlet problem (1), (2) in a different manner than the one just described. This is accomplished by representing $\phi(z)$ in the form

$$\phi(z) = \frac{1}{\pi i}\int_{\partial D}\frac{\mu(t)dt}{t-z}, \quad z\varepsilon D \tag{11}$$

where $\mu(t)$ is a real valued function to be determined, and then using (3) and the limit properties of Cauchy integrals to derive the following integral equation for the unkown potential $\mu(t)$:

$$\mu(t_o) + \int_{\partial D}\left[\mathrm{Re}\,\frac{t'}{\pi i(t-t_o)}\right]\mu(t)ds = f(t_o) \tag{12}$$

where s denotes arclength, $t' = {}^{dt}/ds$ and $t_o \varepsilon \partial D$. Equation (12) is a Fredholm integral equation of the second kind for the unknown density $\mu(t)$ and this equation always has a unique solution for $f(t)\varepsilon C^o(\partial D)$ (Note that for any $\varepsilon > 0, |t_o - t|^{\varepsilon}\mathrm{Re}(\frac{t'}{\pi i(t-t_o)})$ satisfies a Hölder condition for $t\varepsilon\partial D$). If one now appeals to various methods for approximating solutions to Fredholm integral equations of the second kind, one is lead to a constructive method for approximating the solution to the Dirichlet problem (1), (2).

A major part of these lectures will be to extend the methods just described to equations with variable coefficients,in particular to

1) Second order elliptic equations in two independent variables.
2) Second order parabolic equations in one space variable.
3) Second order parabolic equations in two space variables.
4) Certain classes of second order elliptic equations in $n \geqslant 2$ independent variables with spherically symmetric coefficients.

The above extension will often be based on the development of methods for the analytic continuation of solutions to elliptic and parabolic equations. (We will for the sake of simplicity often restrict our attention to the case when the coefficients of the partial differential equation under investigation are entire functions of their independent complex variables. In practice this is not a serious restriction since the coefficients are in general obtained from physical measurements and can be approximated on compact sets by polynomials).

In addition to approximating solutions of boundary value problems for partial differential equations by means of a complete family of solutions, or the method of integral equations, we will also be interested in solving various (in general non linear) problems through the use of inverse methods and analytic function theory. The simplest example of such a problem is the inverse Stefan problem for the heat equation in one space variable, which can be formulated as follows. Consider a thin block of ice at $0^{\circ}C$ occupying the interval $0 \leq x < \infty$ and suppose at $x=0$ the temperature is given by a prescribed function $\phi(t) > 0$ where $t \geq 0$ denotes time. Then the ice will begin to melt and for $t>0$ the water will occupy an interval $0 \leq x < s(t)$. If $u(x,t)$ is the temperature of the water we have

$$\frac{k}{\rho c} u_{xx} = u_t \qquad \text{for } 0<x<s(t) \tag{13}$$

$$u(0,t) = \phi(t) \qquad \text{for } t>0 \tag{14}$$

$$u(s(t),t) = 0 \qquad \text{for } t>0 \tag{15}$$

and, from the law of conservation of energy,

$$u_x(s(t),t) = -\frac{\lambda\rho}{k}\frac{ds(t)}{dt}, \tag{16}$$

where λ, k, ρ, and c are thermal constants. $s(t)$ is an unknown free boundary and the Stefan problem is to determine $u(x,t)$ and $s(t)$ from (13)-(16).

It is easily seen that this problem is non linear in $s(t)$. The inverse Stefan problem is, given $s(t)$, to determine $u(x,t)$ for $0<x<s(t)$ and $\phi(t)=u(0,t)$, i.e. now must we heat the water in order to melt the ice along a prescribed curve? The idea is to construct a "catalog" of solutions $u(x,t)$ corresponding to a large class of "free" boundaries $s(t)$ and to then be in a position to solve the Stefan problem (13)-(16) by looking in the "catalog" for a solution whose boundary data at $x=0$ is close to $\phi(t)$. The inverse Stefan problem is linear; however it is <u>improperly posed</u> in the real domain in the sense that $u(x,t)$ does not depend continuously on the initial data on the curve $s(t)$. To see this let $s(t) = 0$ and assume $\frac{k}{\rho c} = 1$. Then

$$u_n(x,t) = \frac{1}{n}\left[e^{nx}\sin(2n^2t+nx)+e^{-nx}\sin(2n^2t-nx)\right] \qquad (17)$$

is a solution of (13) such that

$$u_n(0,t) = \frac{2}{n}\sin 2n^2t \qquad (18)$$

$$u_{nx}(0,t) = 0 \qquad (19)$$

But although $u_n(0,t)\to 0$ as $n\to\infty$, for any $x>0$, $u_n(x,t)\to\infty$ as $n\to\infty$. However, as a consequence of the Cauchy-Kowalewski theorem, the inverse Stefan problem is well posed in the <u>complex</u> domain and this is where we will later study it.

<u>Remark</u>: The Cauchy-Kowalewski theorem does not provide a practical method for solving the inverse Stefan problem, particularly in higher dimensional space, since the calculations are far too tedious and, more important, the solution may not converge in a large enough domain.

The inverse problems we will study in these lectures are

1) Inverse methods for solving boundary value problems arising in the theory of compressible fluid flow.
2) The inverse Stefan problem for the heat equation in one and two space variables.

3) The inverse scattering problem for acoustic waves in a spherically stratified medium.

In order to accomplish the program outlined above we will use the theory of <u>integral operators</u> as developed by Bergman and Vekua for elliptic equations in two independent variables, by Bergman, Colton and Gilbert for ellitpic equations in three independent variables, and by Colton for parabolic equations in one and two space variables. Suitable references in the case of elliptic equations are [2], [6], [8], [30], [31], [49].

<u>Remark</u>: [8] also contains material on parabolic, hyperbolic and pseudoparabolic equations.

I Elliptic equations in two independent variables

1.1 Analytic Continuation

We are interested here in classical solutions of the second order elliptic equation in two independent variables written in canonical form as

$$L[u] \equiv u_{xx}+u_{yy}+a(x,y)u_x+b(x,y)u_y+c(x,y)u=0 \qquad (1.1.1)$$

where we assume that a,b and c are entire functions of their independent complex variables x and y.

Remark: The assumption that a,b and c are entire functions can be easily relaxed to being analytic in a sufficiently large polydisc in $\mathbb{C}^2$, the space of two complex variables.

We will also need to look at special solutions of the adjoint equation to $L[u]=0$ defined by

$$M[v] \equiv v_{xx} + v_{yy} - \frac{\partial(av)}{\partial x} - \frac{\partial(bv)}{\partial y} + cv = 0. \qquad (1.1.2)$$

In particular we first want to construct a special entire solution of (1.1.2) known as the (complex) Riemann function for $L[u]=0$ ([49]).

To this end we define a mapping of $\mathbb{C}^2 \to \mathbb{C}^2$ by

$$\begin{aligned} z &= x + iy \\ z^* &= x - iy . \end{aligned} \qquad (1.1.3)$$

Note that $z^* = \bar{z}$ if and only if x and y are real.

Under the transformation (1.13) equations (1.1.1) and (1.1.2) become

$$L^*[U] \equiv \frac{\partial^2 V}{\partial z \partial z^*} + A(z,z^*)\frac{\partial U}{\partial z} + B(z,z^*)\frac{\partial U}{\partial z^*} + C(z,z^*)U = 0 \qquad (1.1.4)$$

$$M^*[V] \equiv \frac{\partial^2 V}{\partial z \partial z^*} - \frac{\partial(AV)}{\partial z} - \frac{\partial(BV)}{\partial z^*} + CV = 0 \qquad (1.1.5)$$

where

$$U(z,z^*) = U\left(\frac{z+z^*}{2}, \frac{z-z^*}{2i} \right)$$

$$V(z,z^*) = v\left(\frac{z+z^*}{2}, \frac{z-z^*}{2i} \right)$$

$$A = \frac{1}{4}(a+ib) \tag{1.1.6}$$

$$B = \frac{1}{4}(a-ib)$$

$$C = \frac{1}{4} c$$

and we are assuming that $u(x,y)$ and $v(x,y)$ are analytic functions of the complex variables x and y. We will show later that this is true for every classical solution of (1.1.1) or (1.1.2).

The Riemann function for (1.1.1) is defined to be the (unique) solution $R(z,z^*;\zeta,\zeta^*)$ of (1.1.5) depending on the complex parameters $\zeta = \xi + i\eta$, $\xi^* = \xi - i\eta$ (where ξ,η are complex variables) which satisfies the initial conditions

$$R(z,\zeta^*;\zeta,\zeta^*) = \exp\left[\int_{\zeta}^{z} B(\sigma,\zeta^*)d\sigma \right] \tag{1.1.7}$$

$$R(\zeta,z^*;\zeta,\zeta^*) = \exp\left[\int_{\zeta^*}^{z^*} A(\zeta,\tau)d\tau \right]$$

on the complex hyperplanes $z^* = \zeta^*$ and $z = \zeta$. Note that by Cauchy's theorem the integrals in (1.1.7) are independent of the path of integration. We will now construct $R(z,z^*;\zeta,\zeta^*)$. (1.1.5) and (1.1.7) are equivalent to the integral equation

$$\begin{aligned} R(z,z^*;\zeta,\zeta^*) &- \int_{\zeta}^{z} B(\sigma,z^*)R(\sigma,z^*;\zeta,\zeta^*)d\sigma \\ &- \int_{\zeta^*}^{z^*} A(z,\tau)R(z,\tau;\zeta,\zeta^*)d\tau \\ &+ \int_{\zeta}^{z}\int_{\zeta^*}^{z^*} C(\sigma,\tau)R(\sigma,\tau;\zeta,\zeta^*)d\tau d\sigma = 1 \end{aligned} \tag{1.1.8}$$

We will show that there exists a solution of (1.1.8) which is an entire function of z, z^*, ζ and ζ^*. It suffices to show that $R(z,z^*;\zeta,\zeta^*)$ is an analytic function of its independent variables for $|z-\zeta| < r$ and $|z^*-\zeta^*| < r$ for $r > 1$ an arbitrary large positive number.

We define the recursive scheme

$$R_o(z,z^*) = 0$$

$$\begin{aligned} R_{n+1}(z,z^*) = 1 &+ \int_\zeta^z B(\sigma,z^*)R_n(\sigma,z^*)d\sigma \\ &+ \int_{\zeta^*}^{z^*} A(z,\tau)R_n(z,\tau)d\tau \\ &- \int_\zeta^z\int_{\zeta^*}^{z^*} C(\sigma,\tau)R_n(\sigma,\tau)d\tau d\sigma, \quad n \geqslant 1 \end{aligned} \tag{1.1.9}$$

where $R_n(z,z^*) = R_n(z,z^*;\zeta,\zeta^*)$ and will show that

$$R = \lim_{n\to\infty} R_n = \sum_{n=0}^{\infty} (R_{n+1} - R_n)$$

converges absolutely and uniformly for $|z-\zeta| \leqslant \frac{r}{\alpha}$, $|z^*-\zeta^*| \leqslant \frac{r}{\alpha}$, $\alpha > 1$ arbitrary, and, since each R_n is analytic for $|z-\zeta| < r$, $|z^*-\zeta^*| < r$, so is the limit R. To this end we make use of the <u>method of dominants</u>. If we are given two series

$$S = \sum_{n=0}^{\infty} a_n z^n \quad , \quad \tilde{S} = \sum_{n=0}^{\infty} \tilde{a}_n z^n \quad ; \quad |z| < r$$

where $\tilde{a}_n \geqslant 0$ then we say $\tilde{S}$ dominates S if $|a_n| \leqslant \tilde{a}_n$, $n = 0,1,2, \ldots$, and write $S << \tilde{S}$. It is easily verified that dominates can be multiplied and if $S << \tilde{S}$ then

$$\begin{aligned} &1) \quad \frac{\partial S}{\partial z} << \frac{\partial \tilde{S}}{\partial z} \\ &2) \quad \int_o^z S(z)dz << \int_o^z \tilde{S}(z)dz \\ &3) \quad S << \frac{\tilde{S}}{(1-az)}, \quad 0 \leqslant a < \frac{1}{r}. \end{aligned} \tag{1.1.10}$$

The generalization of the above definition to series not expanded about the origin and to series of several complex variables is immediate.

From (1.1.9) we have, setting $u_n = R_{n+1} - R_n$,

$$u_{n+1}(z,z^*) = \int_{\zeta}^{z} B(\sigma,z^*)u_n(\sigma,z^*)d\sigma + \int_{\zeta^*}^{z^*} A(z,\tau)u_n(z,\tau)d\tau \tag{1.1.11}$$

$$- \int_{\zeta}^{z}\int_{\zeta^*}^{z^*} C(\sigma,\tau)u_n(\sigma,\tau)d\tau d\sigma$$

with $u_0 = 1$. Let M be a positive constant such that for $|z-\zeta| < r$ and $|z^*-\zeta^*| < r$

$$A(z,z^*) << \frac{M}{(1-\frac{z-\zeta}{r})(1-\frac{z^*-\zeta^*}{r})}$$

$$B(z,z^*) << \frac{M}{(1-\frac{z-\zeta}{r})(1-\frac{z^*-\zeta^*}{r})} \tag{1.1.12}$$

$$C(z,z^*) << \frac{M}{(1-\frac{z-\zeta}{r})(1-\frac{z^*-\zeta^*}{r})}$$

where M can be chosen independent of ζ and ζ^* for ζ and ζ^* bounded.

We claim that for $|z-\zeta| < r$, $|z^*-\zeta^*| < r$, $n \geqslant 1$

$$u_n(z,z^*) << \frac{3^n M^n r^{2n}}{(n-1)!} (1- \frac{(z-\zeta)}{r})^{-n} (1- \frac{(z^*-\zeta^*)}{r})^{-n} . \tag{1.1.13}$$

From the properties of dominants and (1.1.11) it is seen that this is clearly true for n = 1. We will now establish (1.1.13) by induction. Assume (1.1.13) is true for n = k.

Then

$$u_{k+1} << \frac{3^k M^{k+1} r^{2k+2}}{k!} \Big[\frac{1}{r}(1- \frac{(z-\zeta)}{r})^{-k} (1- \frac{(z^*-\zeta^*)}{r})^{-k-1}$$

$$+ \frac{1}{r}(1- \frac{(z-\zeta)}{r})^{-k-1} (1- \frac{(z^*-\zeta^*)}{r})^{-k} +$$

$$+\frac{1}{k}\left(1-\frac{(z-\zeta)}{r}\right)^{-k-1}\left(1-\frac{(z^*-\zeta^*)}{r}\right)^{-k-1}\Big] \tag{1.1.14}$$

$$< \frac{3^{k+1}M^{k+1}r^{2k+2}}{k!}\left(1-\frac{(z-\zeta)}{r}\right)^{-k-1}\left(1-\frac{(z^*-\zeta^*)}{r}\right)^{-k-1}$$

thus showing (1.1.13) is true for $n = k + 1$ and completing the induction proof. Now let $\alpha > 1$ and $|z-\zeta| \leq \frac{r}{\alpha}$, $|z^*-\zeta^*| \leq \frac{r}{\alpha}$.

Then $\left(1-\frac{|z-\zeta|}{r}\right) \geq \frac{\alpha-1}{\alpha}$, $\left(1-\frac{|z^*-\zeta^*|}{r}\right) \geq \frac{\alpha-1}{\alpha}$, and the series $\sum_{n=0}^{\infty} |u_n| = \sum_{n=0}^{\infty} |R_{n+1}-R_n|$ is majorised by

$$1 + \sum_{n=1}^{\infty} \frac{3^n M^n r^{2n}}{(n-1)!}\left(\frac{\alpha}{\alpha-1}\right)^{2n} < \infty \tag{1.1.15}$$

Hence we have shown that $R(z,z^*;\zeta,\zeta^*)$ exists and is an entire function of its independent variables.

We now want to prove the following theorem:

Theorem 1.1.1 ([49]) : As a function of its last two arguements $R(\zeta,\zeta^*,z,z^*)$ is the Riemann function for $M[v] = 0$.

Proof: Let D be abounded simply connected domain $\mathbb{C}^1$ and let $U(z,z^*)$ be an analytic function of z and z* for $(z,z^*)\varepsilon D\times D^*$ where $D^* = \{z^* : \bar{z}^*\varepsilon D\}$. From $M^*[R] = 0$ and (1.1.7) we have

$$\frac{\partial^2(UR)}{\partial z \partial z} - RL^*[U] = \frac{\partial}{\partial z}\left\{U\left(\frac{\partial R}{\partial z^*} - AR\right)\right\} + \frac{\partial}{\partial z^*}\left\{U\left(\frac{\partial R}{\partial z} - BR\right)\right\} \tag{1.1.16}$$

where $R = R(z,z^*;\zeta,\zeta^*)$. Interchange z,z^* and ζ,ζ^* in (1.1.16) and integrate with respect to ζ and ζ^* from z_o to z and z_o^* to z* where $(z_o,z_o^*)\varepsilon D\times D^*$. Making use of (1.1.7) again we have

$$U(z,z^*) = U(z_o,z_o^*)R(z_o,z_o^*;z,z^*)$$

$$+\int_{z_o}^{z} R(\zeta,z_o^*,z,z^*)\left\{\frac{\partial U(\zeta,z_o^*)}{\partial \zeta} + B(\zeta,z_o^*)U(\zeta,z_o^*)\right\} d\zeta$$

$$+\int_{z_o^*}^{z^*} R(z_o,\zeta^*,z,z^*)\{\frac{\partial U(z_o,\zeta^*)}{\partial\zeta^*} + A(z_o,\zeta^*)U(z_o,\zeta^*)\}d\zeta^*$$

$$+\int_{z_o}^{z}\int_{z_o^*}^{z^*} R(\zeta,\zeta^*,z,z^*)L^*[U(\zeta,\zeta^*)]d\zeta^*d\zeta \tag{1.1.17}$$

Setting $U(z,z^*) = R(z_o,z_o^*,z,z^*)$ and using (1.1.7) shows that

$$\int_{z_o}^{z}\int_{z_o^*}^{z} R(\zeta,\zeta^*;z,z^*)L^*[R(z\ ,z_o^*,\zeta,\zeta^*)]d\zeta^*d\zeta = 0 \tag{1.1.18}$$

i.e. with respect to its last two arguments $R(z_o,z_o^*,z,z^*)$ is a solution of $L^*[U] = 0$. (1.1.7) now shows that a function of z,z^*, $R(\zeta,\zeta^*, z,z^*)$ is the Riemann function for $M[v] = 0$.

Corollary 1.1.1 : Let $F(z,z^*)$ be analytic for $(z,z^*)\varepsilon DxD^*$. Then

$$U_o(z,z^*) = \int_{z_o}^{z}\int_{z_o^*}^{z^*} R(\zeta,\zeta^*\ z,z^*)F(\zeta,\zeta^*)d\zeta^*d\zeta \tag{1.1.19}$$

is a particular solution of $L^*[U] = F(z,z^*)$ analytic for (z,z^*) DxD^*.

Proof: This follows from (1.1.17).

We now want to prove the main result of this section, the Bergman-Vekua Theorem (c.f. [2], [49]).

Theorem 1.1.2 ([2], [49]): Let $u(x,y)$ be a classical solution of $L[u] = 0$ in D. Then $U(z,\bar{z}) = u(x,y)$ is analytic for $(x,y)\varepsilon D$ and $U(z,z^*) = u(\frac{z+z^*}{2}, \frac{z-z^*}{2i})$ can be analytically continued into the domain DxD^*.

Proof: Without loss of generality assume D has a smooth boundary ∂D and let ν denote the inner normal and s the arclength along ∂D. Then from Green's theorem we have for $u,v\varepsilon C^2(D)\cap C^1(\bar{D})$

$$\iint_D (vL[u] - uM[v]dxdy + \int_{\partial D} H[u,v] = 0 \tag{1.1.20}$$

where $H[u,v] = \{v \frac{\partial u}{\partial \nu} - u \frac{\partial v}{\partial \nu} + (a \frac{\partial x}{\partial \nu} + b \frac{\partial y}{\partial \nu})uv\}\, ds$.

Now let u be a solution of $L[u] = 0$ in D (without loss of generality assume $u \in C^1(\bar{D})$) and set $v = R(z,\bar{z}; \zeta,\bar{\zeta}) \log r$ (where $\zeta = \xi + i\eta$, $\bar{\zeta} = \xi - i\eta$, $r^2 = (z-\zeta)(\bar{z}-\bar{\zeta})$). Let Ω_ε be a small circle about the point (ξ,η) and apply (1.1.20) to the region D/Ω_ε instead of D. Letting $\varepsilon \to 0$ and interchanging the roles of (x,y) and (ξ,η) now gives in a straightforward fashion

$$u(x,y) = \frac{1}{2\pi} \int_{\partial D} H[u, R\log r] - \frac{1}{2\pi} \int\int_D uM[R\log r]\, d\xi d\eta \qquad (1.1.21)$$

where integration over D and ∂D is now with respect to the point (ξ,η), $R = R(\zeta,\bar{\zeta}; z,\bar{z})$, and M is a differential operator with respect to the (ξ,η) variables. Since $M[R] = 0$ we have (with respect to the complex variables (ζ,ζ^*, z,z^*))

$$M^*[R\log r] = 2\, \frac{\partial R/\partial\zeta - BR}{\zeta^* - z^*} + 2\, \frac{\partial R/\partial\zeta^* - AR}{\zeta - z} \qquad (1.1.22)$$

and hence from (1.1.7) we have that $M[R\log r]$ is in fact an entire function of its independent complex variables. Hence the second integral in (1.1.21) can be continued to an entire function of z and z* (replace $\bar{z}$ by z*). The first integral in (1.1.21) can be continued to an analytic function of z and z* for $z \in D$, $z^* \in D^*$ (i.e. for z and z* such that $r \neq 0$). Hence (1.1.21) shows that $U(z,z^*)$ is analytic for $(z,z^*) \in D \times D^*$.

Remark: Note that u(x,y) analytic for $(x,y) \in D$ means that for each point $(x_0,y_0) \in D$ there exists a neighbourhood N of (x_0,y_0) in $\mathbb{C}^2$ such that u(x,y) is analytic in N. Theorem 1.1.2 provides a global analytic continuation as opposed to this local result.

1.2 Integral Operators.

Let $u(x,y)$ be a real valued solution of $L[u] = 0$ in a bounded simply connected domain D. We make the further assumption that, in addition to being entire functions, the coefficients a, b and c are real valued for real values of their arguments. Our aim is to construct an integral operator which maps analytic functions of a single complex variable onto solutions of $L[u] = 0$ (c.f. [2],[49]). Without loss of generality we assume that the origin is an interior point of D.

One such operator is already given to us from the results of section 1.1. From Theorem 1.1.2 we have that $U(z,z^*) = u\left(\frac{z+z^*}{2}, \frac{z-z^*}{2i}\right)$ is analytic for $(z,z^*)\varepsilon D\times D^*$ and hence from (1.1.17) we have

$$U(z,z^*) = \alpha_o R(0,0;\ z,z^*) + \int_o^z f(\zeta)R(\zeta,0,z,z^*)d\zeta + \int_o^{z^*} g(\zeta^*)R(0,\zeta^*,z,z^*)d\zeta^* \qquad (1.2.1)$$

where

$$\alpha_o = U(0,0)$$

$$f(z) = \frac{\partial U(z,0)}{\partial z} + B(z,0)U(z,0) \qquad (1.2.2)$$

$$g(z^*) = \frac{\partial U(0,z^*)}{\partial z^*} + A(0,z^*)U(0,z^*).$$

From Theorem 1.1.2 we have that $f(z)$ and $g(z^*)$ are analytic in D and D^* respectively. Conversely, it is easily seen that if $f(z)$ and $g(z^*)$ are any analytic functions in D and D^* respectively, then (1.2.1) defines a solution of $L[u] = 0$. Note that there we have not made use of the fact that u, a, b and c are real valued. The operator defined by (1.2.1) is known as Vekua's integral operator. For more details see [49].

Theorem 1.2.1 ([2], [49]): Let $u(x,y)$ be a solution of $L[u] = 0$ in D. Then $u(x,y)$ is uniquely determined from the complex Goursat data $U(z,0)$ and $U(0,z^*)$.

Proof: If $U(z,0) = U(0,z^*) = 0$ then $\alpha_0 = f(z) = g(z) = 0$ and hence from (1.2.1) $U(z,z^*) \equiv 0$.

Corollary 1.2.1 ([2], [49]): Let $u(x,y)$ be a real valued solution of $L[u] = 0$ in D. Then $u(x,y)$ is uniquely determined by $U(z,0)$.

Proof: From Theorem 1.1.2 we have that in some ball in $\mathbb{C}^2$ about the origin

$$U(z,z^*) = \sum_{m,n=0}^{\infty} a_{mn} z^m z^{*n}. \tag{1.2.3}$$

Since $u(x,y)$ is real valued $U(z,\bar{z}) = \overline{U(z,\bar{z})}$, i.e.

$$\sum_{m,n=0}^{\infty} a_{mn} z^m \bar{z}^n = \sum_{m,n=0}^{\infty} \overline{a_{mn}}\, \bar{z}^m z^n \tag{1.2.4}$$

and hence

$$a_{mn} = \overline{a_{nm}} .$$

Since

$$U(z,0) = \sum_{m=0}^{\infty} a_{mo} z^m$$

$$U(0.z^*) = \sum_{n=0}^{\infty} a_{on} z^{*n} \tag{1.2.5}$$

we have that $U(z,0)$ determines $U(0,z^*)$ and the corollary now follows from Theorem 1.2.1.

We will return later to further discussion of Vekua's integral operator.

We now want to construct another operator which maps analytic functions on to solutions of $L[u] = 0$, the so called Begman integral operator of the first kind (c.f. [2]). We want to do this since it is the Bergman operator which provides the proper motivation for generalizing the method of integral operators to certain classes of elliptic equations in more than two independent variables and to parabolic equations in one and two space variables. In order to construct this operator we will need the assumption that a,b and c are real valued, and we will assume this from now on.

We consider $L[u] = 0$ in its complex form

$$L^*[U] \equiv U_{zz^*} + A(z,z^*)U_z + B(z,z^*)U_z + C(z,z^*)U = 0 \qquad (1.2.6)$$

and make the change of dependent variables

$$V(z,z^*) = U(z,z^*)\exp\left\{ \int_0^{z^*} A(z,\zeta^*)d\zeta^* \right\}. \qquad (1.2.7)$$

Under the change of variables (1.2.7), (1.2.6) becomes

$$V_{zz^*} + D(z,z^*)V_{z^*} + F(z,z^*)V = 0 \qquad (1.2.8)$$

where

$$D = B - \int_0^{z^*} A_z(z,\zeta^*)d\zeta^*$$
$$F = -(A_z + AB - C). \qquad (1.2.9)$$

We look for solutions of (1.2.7) in the form

$$V(z,z^*) = \int_{-1}^{1} E(z,z^*,t)f\left(\frac{z}{2}(1-t^2)\right)\frac{dt}{\sqrt{1-t^2}} \qquad (1.2.10)$$

where f(z) is an analytic function in some neighbourhood of the origin and E(z,z*,t) is to be determined.

<u>Definition 1.2.1</u>: E(z,z*,t) is known as the <u>generating function</u> for equation (1.2.8).

<u>Remark</u>: The path of integration in (1.2.10) is assumed to be a curvilinear path in the unit disc in the complex t plane joining the points t = +1 and t = -1.

Assuming that E(z,z*,t) is an analatyic function of t for $|t| \leq 1$ and (z,z*) in some neighbourhood of the origin in $\mathbb{C}^2$ we substitute (1.2.10) into the differential equation (1.2.8) and integrate by parts using

$$f_z = - f_t \frac{(1-t^2)}{2zt} \tag{1.2.11}$$

to show that if $E(z,z^*,t)$ satisfies

$$(1-t\ )E_{z^*t} - \frac{1}{t} E_{z^*} + 2tz \left[E_{zz^*} + DE_{z^*} + FE \right] = 0 \tag{1.2.12}$$

then (1.2.10) yields a solution of (1.2.8).

We will now show the existence and regularity of $E(z,z^*,t)$. We look for a solution of (1.2.12) in the form

$$E(z,z^*,t) = 1 + \sum_{n=1}^{\infty} t^{2n} z^n \int_0^{z^*} P^{(2n)}(z,z^*)dz^* \quad . \tag{1.2.13}$$

Substituting (1.2.13) into (1.2.12) gives the following recursion formula for the $P^{(2n)}$:

$$P^{(2)} = - 2F \tag{1.2.14}$$

$$(2n+1)P^{(2n+2)} = -2\left[P_z^{(2n)} + DP^{(2n)} + F \int_0^{z^*} P^{(2n)} dz^* \right] \ ; \ n \geq 1 \ .$$

Note that since D and F are entire functions of z and z^*, so are the $P^{(2n)}$. We will now show that the series (1.2.13) converges absolutely and uniformly for (t,z,z^*) on compact subsets of $\mathbb{C}^3$. To do this we will again use the method of dominants. Let r be an arbitrarily large positive number and M a positive constant such that for $|z| < r$, $|z^*| < r$, we have

$$D(z,z^*) << \frac{M}{(1- \frac{z}{r})(1- \frac{z^*}{r})} \tag{1.2.15}$$

$$F(z,z^*) << \frac{M}{(1- \frac{z}{r})(1- \frac{z^*}{r})} \quad .$$

We will now show by induction that there exist positive constants M_n and ε (where ε is independent of n and M_n is a bounded function of n) such that for $|z| < r$, $|z^*| < r$, we have

$$P^{(2n)} << \frac{M_n 2^n (1+\varepsilon)^n}{2n-1} (1-\frac{z}{r})^{-(2n-1)} (1-\frac{z^*}{r})^{-(2n-1)} r^{-n} . \qquad (1.2.16)$$

This is clearly true for n = 1. Now suppose for n = k (1.2.16)is valid. Then from (1.2.14) and (1.2.15) and the straightforward use of the method of dominants we have

$$P^{(2k+2)} << \frac{M_k 2^{k+1} (1+\varepsilon)^k}{(2k+1)} \left[1+ \frac{Mr}{2k-1} + \frac{Mr^2}{(2k-1)(2k-1)} \right] .$$

$$.(1-\frac{z}{r})^{-(2k+1)} (1-\frac{z^*}{r})^{-(2k+1)} r^{-k-1} . \qquad (1.2.17)$$

By setting

$$M_{n+1} = M_n (1+\varepsilon)^{-1} \{1+\varepsilon \frac{Mr}{2n-1} + \frac{Mr^2}{(2n-1)(2n-1)}\} \qquad (1.2.18)$$

$$M_1 = M$$

we have shown that (1.2.16) is true for n = k + 1, thus completing the induction step. Note that for n sufficiently large we have $M_{n+1} \leqslant M_n$, i.e. there exists a positive constant M_o which is independent of n such that $M_n \leqslant M_o$ for all n.

We now return to the convergence of (1.2.13). Let $t_o \geqslant 1$ and $\alpha > 1$ be positive constants and let $|t| \leqslant t_o$, $|z| < \frac{r}{\alpha}$, $|z^*| < \frac{r}{\alpha}$. Then $(1-\frac{|z|}{r}) \geqslant \frac{\alpha-1}{\alpha}$, $(1-\frac{|z^*|}{r}) \geqslant \frac{\alpha-1}{\alpha}$, and from (1.2.16) it is seen that the series (1.2.13) is majorised by

$$1 + \sum_{n=1}^{\infty} \frac{rM_n 2^n t_o{}^n (1+\varepsilon)^n \alpha^{3n-3}}{(2n-1)(\alpha-1)^{4n-2}} . \qquad (1.2.19)$$

If α is chosen such that

$$2\alpha^3 t_o(1+\varepsilon)(\alpha-1)^{-4} < 1 , \qquad (1.2.20)$$

then the series (1.2.19) is convergent. Since r is an arbitrarily large positive number and ε is arbitrarily small and independent of r, we can now conclude that the series (1.2.13) converges absolutely and uniformly on compact subsets of $\mathbb{C}^3$, i.e. $E(z.z^*,t)$ is an entire function of its independent complex variables.

We have now shown that the operator $\underset{\sim}{B}_2$ defined by

$$\begin{aligned} U(z,z^*) &= \underset{\sim}{B}_2\{f\} \\ &= \exp \{ - \int_0^{z^*} A(z,\zeta^*)d\zeta^* \} . \qquad (1.2.21) \\ &\cdot \int_{-1}^{1} E(z,z^*,t)f(\tfrac{z}{2}(1-t^2)) \frac{dt}{\sqrt{1-t^2}} \end{aligned}$$

exists and maps analytic functions which are regular in some neighbourhood of the origin in $\mathbb{C}^1$ <u>into</u> the class of (complex valued) solutions of $L^*[U] = 0$. We now make use of Corollary 1.2.1 to show that the operator Re $\underset{\sim}{B}_2$, where "Re" denotes "take the real part", maps analytic functions <u>onto</u> the class of real valued solutions of $L[u] = 0$. We first note that since the coefficients of $L[u] = 0$ are real valued for x and y real, Re $\underset{\sim}{B}_2\{f\}$ defines a real valued solution of $L[u] = 0$ provided we set $z^* = \bar{z}$. Evaluating Re $\underset{\sim}{B}_2\{f\}$ at $z^* = 0$ gives

$$\begin{aligned} U(z,0) &= (\text{Re } \underset{\sim}{B}_2\{f\})_{z^*=0} \\ &= \frac{1}{2}\int_{-1}^{1} [f(\tfrac{z}{2}(1-t^2))+\bar{f}(0)\exp(-\int_0^z \bar{A}(\sigma,\zeta)d\zeta)] \frac{dt}{\sqrt{1-t^2}} \qquad (1.2.22) \\ &= \frac{1}{2}\int_{-1}^{1} f(\tfrac{z}{2}(1-t^2)) \frac{dt}{\sqrt{1-t^2}} + \frac{\pi}{2}\bar{f}(0)\exp(-\int_0^z \bar{A}(0,\zeta)d\zeta) \end{aligned}$$

where $\bar{f}(z) = \overline{f(\bar{z})}$ and $\bar{A}(z,z^*) = \overline{A(\bar{z},\bar{z}^*)}$. From (1.2.22) we have $U(0,0) = \frac{\pi}{2}(f(0) + \bar{f}(0))$ and so without loss of generality we can assume f(0) is real and $f(0) = \frac{1}{\pi} U(0,0)$. Hence to show the invertibility of Re $\underset{\sim}{B}_2$ it follows from (1.2.22) and Corollary 1.2.1 that we must be able to invert the integral equation

$$g(z) = \int_{-1}^{1} f\left(\frac{z}{2}(1-t^2)\right) \frac{dt}{\sqrt{1-t^2}} . \tag{1.2.23}$$

Setting

$$g(z) = \sum_{n=0}^{\infty} a_n z^n, \qquad |z| < \rho \tag{1.2.24}$$

it follows from the definition of the Gamma function that

$$-\frac{1}{2\pi} \int_{-1}^{1} g(z(1-t^2)) \frac{dt}{t^2} = \sum_{n=0}^{\infty} \frac{\Gamma(n+1) a_n z^n}{\Gamma(\frac{1}{2})\Gamma(n+\frac{1}{2})} \tag{1.2.25}$$

(where in (1.2.25) the path of integration does not pass through the origin) and, setting $f(^z/2)$ equal to the right hand side of (1.2.25), that

$$\int_{-1}^{1} f\left(\frac{z}{2}(1-t^2)\right) \frac{dt}{\sqrt{1-t^2}} \tag{12.26}$$

$$= \sum_{n=0}^{\infty} \frac{\Gamma(n+1) a_n z^n}{\Gamma(\frac{1}{2})\Gamma(n+\frac{1}{2})} \int_{-1}^{1} (1-t^2)^{n-\frac{1}{2}} dt$$

$$= \sum_{n=0}^{\infty} a_n z^n$$

$$= g(z).$$

Summarizing the above results gives the following theorem:

<u>Theorem 1.2.2</u> ([2]): Let $u(x,y)$ be a real valued (classical) solution of $L[u] = 0$ in some neighbourhood of the origin. Then $u(x,y) = U(z,\bar{z})$ can be represented in the form $U(z,\bar{z}) = \text{Re } \underset{\sim}{B}_2\{f\}$ where $f(z)$ is analytic in some neighbourhood of the origin in $\mathbb{C}^1$. Conversely, for every analytic function $f(z)$ defined in some neighbourhood of the origin in $\mathbb{C}^1$, Re$\underset{\sim}{B}_2\{f\}$ defines a real valued solution of $L[u] = 0$ in some neighbourhood of the origin.

Corollary 1.2.2 ([2]): Let $u(x,y) = \operatorname{Re} \underset{\sim}{B}_2\{f\}$ and suppose $u(x,y)$ is regular in D (i.e. $u(x,y) \varepsilon C^2(D)$). Then $f(\frac{z}{2})$ is analytic for $z = x+iy \varepsilon D$.

Proof: From Theorem 1.1.2 we have that

$$g(z) = 2U(z,0) - U(0,0)\exp(-\int_0^z \bar{A}(0,\zeta)d\zeta) \tag{1.2.27}$$

is analytic for $z\varepsilon D$. From (1.2.22), (1.2.25) we have

$$f(\frac{z}{2}) = -\frac{1}{2\pi}\int_{-1}^{1} g(z(1-t^2))\frac{dt}{t^2} \tag{1.2.28}$$

and, by deforming the path of integration in (1.2.28) if necessary, it is seen that $f(\frac{z}{2})$ is also analytic for $z\varepsilon D$.

1.3. Complete Families of Solutions.

In this section we will make the further assumption on D, that in addition to being bounded and simply connected, D is in class Ah, i.e. the angle $\theta(t)$ between the tangent to ∂D at the point t and the x axis is Hölder continuous along ∂D. Without loss of generality we assume that D contains the origin. We want to construct a set of solutions $\{u_n\}$ to $L[u] = 0$ such that if $u(x,y)\varepsilon C^0(\bar{D}) \cap C^2(D)$ is a real valued solution of $L[u] = 0$ in D then for any $\varepsilon > 0$ there exists an integer $N = N(\varepsilon)$ and constants $a_0, \ldots, a_N$ such that

$$\max_{\bar{D}} \left| u - \sum_{n=0}^{N} a_n u_n \right| < \varepsilon .$$

Then set $\{u_n\}$ is then said to be complete in the maximum norm over $\bar{D}$. From Runge's theorem for analytic functions, Corollary 1.2.2, and the regularity of $E(z,z^*,t)$ we immediately have the following theorem:

Theorem 1.3.1 ([2] [49]): Let $u(x,y)\varepsilon C^2(D)$ be a real valued solution of $L[u] = 0$ in D and let

$$\begin{aligned} u_{2n} &= \operatorname{Re} \underset{\sim}{B}_2\{z^n\} \quad ; \quad n = 0,1, \ldots \\ u_{2n+1} &= \operatorname{Im} \underset{\sim}{B}_2\{z^n\}; \quad n = 0,1, \ldots \end{aligned} \tag{1.3.1}$$

where "Re" denotes "take the real part" and "Im" denotes "take the imaginary part". Let D_0 be a compact subset of D. Then for any $\varepsilon > 0$ there exists an integer $N = N(\varepsilon)$ and constants $a_0, \ldots, a_N$ such that

$$\max_{\bar{D}_0} \left|u - \sum_{n=0}^{N} a_n u_n\right| < \varepsilon \; . \tag{1.3.2}$$

The problem therefore is to replace D_0 in (1.3.2) by $\bar{D}$, i.e. to show the set $\{u_n\}$ is "complete up to the boundary". We will show this through the use of singular integral equations and the method of I.N.Vekua ([49]).

We will first need a few preliminary results concerning the elliptic equation in the complex domain

$$L^*[U] \equiv \frac{\partial^2 U}{\partial z \partial z^*} + A(z,z^*)\frac{\partial U}{\partial z} + B(z,z^*)\frac{\partial U}{\partial z^*} + C(z,z^*)U = 0 \; . \tag{1.3.3}$$

Let

$$\Lambda(z,z^*) = \exp\{-\int_0^z B(\zeta,0)d\zeta - \int_0^{z^*} A(0,\zeta^*)d\zeta^* \tag{1.3.4}$$
$$+ \int_0^z \int_0^{z^*} [A(\zeta,\zeta^*)B(\zeta,\zeta^*) - C(\zeta,\zeta^*)]d\zeta^* d\zeta\}$$

and set

$$U = \Lambda U' \; . \tag{1.3.5}$$

Then $U'(z,z^*)$ satisfies

$$L'[U'] \equiv \frac{\partial^2 U'}{\partial z \partial z^*} + A'(z,z^*)\frac{\partial U'}{\partial z} + B'(z,z^*)\frac{\partial U'}{\partial z^*} + A'(z,z^*)B'(z,z^*)U' = 0 \tag{1.3.6}$$

where

$$A'(z,z^*) = \int_0^z h(\zeta,z^*)d\zeta \tag{1.3.7}$$
$$B'(z,z^*) = \int_0^z k(z,\zeta^*)d\zeta^*$$

with

$$h(z,z^*) = \frac{\partial A(z,z^*)}{\partial z^*} + A(z,z^*)B(z,z^*) - C(z,z^*) \tag{1.3.8}$$
$$k(z,z^*) = \frac{\partial B(z,z^*)}{\partial z^*} + A(z,z^*)B(z,z^*) - C(z,z^*) \; ,$$

i.e.

$$A'(0,z^*) = B'(z,0) = 0 \quad . \tag{1.3.9}$$

Hence we can assume that the coefficients $A(z,z^*)$ and $B(z,z^*)$ satisfy (1.3.9) to begin with.

Lemma 1.3.1 ([49]): The Riemann function $R(\zeta,\zeta^* : z,z^*)$ of $L[u] = 0$ with real-valued coefficients takes real values when $z^* = \bar{z}$, $\zeta^* = \bar{\zeta}$.

Proof: Since the coefficients of $L[u] = 0$ are real we have

$$\text{Im } R(\zeta,\bar{\zeta} \; ; \; z,\bar{z}) = \frac{1}{2i}\left[R(\zeta,\bar{\zeta} \; ; \; z,\bar{z}) - \bar{R}(\bar{\zeta},\zeta : \bar{z},z)\right] \tag{1.3.10}$$

where $\bar{R}(\zeta,\bar{\zeta} \; ; \; z,\bar{z}) = \overline{R(\bar{\zeta},\zeta \; ; \; \bar{z},z)}$ is a solution of $L[u] = 0$. Extending (1.3.10) into the complex domain and evaluating along the characteristic $z=0$ gives

$$\begin{aligned} \text{Im } R(\zeta,\bar{\zeta} \; ; \; 0,z^*) &= \frac{1}{2i} \left[\exp\left(-\int_{\bar{\zeta}}^{z^*} A(\zeta,\tau)d\tau\right)\right. \\ &\left. - \exp\left(-\int_{\bar{\zeta}}^{z^*} \bar{B}(\sigma,\tau)d\sigma\right) \right] \\ &= 0 \end{aligned} \tag{1.3.11}$$

from (1.3.7) and the fact that $A(z,\bar{z}) = \overline{B(z,\bar{z})}$ (since the coefficients of $L[u] = 0$ are real). Similarly

$$\text{Im } R(\zeta,\bar{\zeta} \; ; \; z,0) = 0 \; , \tag{1.3.12}$$

and hence from Theorem 1.2.1 $\text{Im } R(\zeta,\bar{\zeta} \; ; \; z,\bar{z}) \equiv 0$ and the theorem is proved.

For the remainder of this section we assume that the coefficients of (1.3.3) satisfy (1.3.9) and that $L[u] = 0$ has real-valued coefficients.

Lemma 1.3.2. ([49]): Let $u(x,y)$ be a real valued solution of $L[u] = 0$ in D. Then

$$u(x,y) = \text{Re}\left[H_0(z)\phi(z) + \int_0^z H(z,\zeta)\phi(\zeta)d\zeta\right] \tag{1.3.13}$$

where

$$H_o(z) = R(z,0,z,\bar{z})$$

$$H(z,\zeta) = -\frac{\partial}{\partial\zeta} R(\zeta,0\ ;\ z,\bar{z}) \tag{1.3.14}$$

$$\phi(z) = 2U(z,0) - U(0,0)$$

and, as usual, $U(z,z^*) = u\left(\frac{z+z^*}{2}, \frac{z-z^*}{2i}\right)$.

Proof: Integrating by parts in (1.2.1) gives, using (1.3.9),

$$\begin{aligned} U(z,z^*) &= -U(0,0)R(0,0\ ;\ z,z^*) \\ &+U(z,0)R(z,0;z,z^*)-\int_0^z U(\zeta,0)\frac{\partial R}{\partial\zeta}(\zeta,0;z,z^*)d\zeta \\ &+U(0,z^*)R(0,z^*;z,z^*)-\int_0^{z} U(0,\zeta^*)\frac{\partial R}{\partial\zeta}(0,\zeta^*;z,z^*)d\zeta^* \end{aligned} \tag{1.3.15}$$

and hence from (1.1.7) and (1.3.9)

$$\begin{aligned} U(z,z^*) &= R(z,0;z,z^*)\Phi(z) - \int_0^z \Phi(\zeta)\frac{\partial R}{\partial\zeta}(\zeta,0;z,z^*)d\zeta \\ &+R(0,z^*;z,z^*)\Phi(z^*) - \int_0^{z^*} \Phi(\zeta^*)\frac{\partial R}{\partial\zeta^*}(0,\zeta^*;z,z^*)d\zeta^* \end{aligned} \tag{1.3.16}$$

where

$$\begin{aligned} \Phi(z) &= U(z,0) - \tfrac{1}{2}U(0,0) \\ \Phi(z^*) &= U(0,z^*) - \tfrac{1}{2}U(0,0). \end{aligned} \tag{1.3.17}$$

Since $u(x,y)$ is real valued we have $U(z,\bar{z}) = \overline{U(z,\bar{z})}$ and $\Phi(z) = \overline{\Phi(\bar{z})}$. Hence from lemma 1.3.1 we can write (1.3.16) as (1.3.13) with $\phi(z)$ as given in (1.3.14).

Now assume that $u(x,y)\varepsilon C^o(\bar{D})$ is a solution of $L[u] = 0$ in D such that $u(t) = f(t)$ for $t\varepsilon\partial D$ where $f(t)$ is Hölder continuous on ∂D. Let $\mu(t)$ be an real valued Hölder continuous function for $t\varepsilon\partial D$. We will try and determine $\mu(t)$ such that $u(x,y)$ can be represented in the form (1.3.13) with

$$\phi(z) = \int_{\partial D} \frac{t\mu(t)ds}{t-z} \tag{1.3.18}$$

where ds denotes arclength along ∂D. If such a $\mu(t)$ exists then we can conclude that $\phi(z)$ as given by (1.3.18) is Hölder continuous in $\bar{D}$; c.f. [42]. Substituting (1.3.18) into (1.3.13) gives

$$u(x,y) = \int_{\partial D} K(z,t)\mu(t)ds \tag{1.3.19}$$

where

$$K(z,t) = \mathrm{Re}\left[\frac{tH_o(z)}{t-z} + \int_o^z \frac{tH(z,t_1)}{t-t_1}\,dt_1\right] \tag{1.3.20}$$

and $t\varepsilon\partial D$, $z\varepsilon D$. Note that $K(z,t)$ has the form

$$K(z,t) = \mathrm{Re}\left[\frac{tH_o(z)}{t-z} - tH(z,t)\log(1-\frac{z}{t}) + H^*(z,t)\right] \tag{1.3.21}$$

where

$$H^*(z,t) = \int_o^z \frac{t\left[H(z,t_1)-H(z,t)\right]}{t-t_1}\,dt_1 \tag{1.3.22}$$

is an analytic function of z,t in DxD and $\log(1-\frac{z}{t})$ is understood to be its principal value. Note also that for fixed $t\varepsilon\partial D$, $K(z,t)$ is a solution of $L[u] = 0$ in D.

Now let $z = xiy\varepsilon D$ tend to a point $t_o\varepsilon\partial D$. From the limit properties of Cauchy integrals (c.f. [42]) we have

$$A(t_o)\mu(t_o) + \int_{\partial D} K(t_o,t)\mu(t)ds = f(t_o) \tag{1.3.23}$$

where

$$\begin{aligned} A(t_o) &= \mathrm{Re}\left[i\pi t_o\bar{t}_o{}'H_o(t_o)\right] \\ K(t_o,t) &= \mathrm{Re}\left[\frac{tH_o(t_o)}{t-t_o} - tH(t_o,t)\log(1-\frac{t_o}{t}) + H^*(t_o,t)\right] \end{aligned} \tag{1.3.24}$$

and $t' = {}^{dt}/ds = e^{i\theta(t)}$ where $\theta(t)$ is the angle between the positive direction of the tangent to ∂D at the point t to the x axis. Since D is in class Ah we have that $t'(s)$ is Hölder continuous on ∂D. (1.3.23) is a singular integral equation for the unknown function $\mu(t)$. We now rewrite (1.3.23) in the form

$$A(t_0)\mu(t_0) + \frac{B(t_0)}{i\pi}\int_{\partial D}\frac{\mu(t)dt}{t-t_0} + \int_{\partial D} K_0(t_0,t)\mu(t)ds = f(t_0) \qquad (1.3.25)$$

where

$$B(t_0) = i\pi \operatorname{Re}\left[t_0\bar{t}_0{}'H_0(t_0)\right] \qquad (1.3.26)$$

$$K_0(t_0,t) = K(t_0,t) - \frac{t'B(t_0)}{i\pi(t-t_0)}$$

and note that $K_0(t_0,t)$ has the form

$$K_0(t_0,t) = \frac{K^*(t_0,t)}{|t-t_0|^\alpha} \qquad (1.3.27)$$

where $0 \leqslant \alpha < 1$ and $K^*(t_0,t)$ is a Hölder continuous function on $\partial D \times \partial D$.

From (1.3.24) and (1.3.26) we have

$$A(t_0) + B(t_0) = i\pi t_0\overline{t_0{}'}H_0(t_o)$$
$$A(t_o) - B(t_o) = -\,i\pi\bar{t}_o t_o{}'\overline{H_o(t_o)} \qquad (1.3.28)$$

and since $H_0(t_0) \neq 0$ for $t_0 \in \partial D$ we have that $A+B$ and $A-B$ are nonzero on ∂D. Hence (1.3.23) is of <u>normal type</u> and the general theory of singular integral equations can be applied (c.f. [42]).

From (1.3.28) we have that the <u>index</u> κ of (1.3.23) is

$$\kappa = \frac{1}{2\pi i}\left[\log\frac{A(t_0)-B(t_0)}{A(t_0)+A(t_0)}\right]_{\partial D} = 0 \qquad (1.3.29)$$

(this follows from the facts that $[\log t_0\bar{t}_0{}']_{\partial D} = [\log \bar{t}_0 t_0{}']_{\partial D} = 0$ and $H_0(t_0) = \exp(-\int_0^{\bar{t}} A(t_0,\eta)d\eta)$, which implies $[\log H_0(t_0)]_{\partial D} = [\log \overline{H_0(t_0)}]_{\partial D} =$ and hence all three Fredholm theorems hold for equation (1.3.23), in particula (1.3.23) has a unique Hölder continuous solution for any (Hölder continuous) $f(t)$ if and only if the homogeneous equation

$$A(t_o)\mu(t_o) + \int_{\partial D} K(t_o,t)\mu(t)ds = 0 \qquad (1.3.30)$$

has only the trivial solution $\mu(t) = 0$ on ∂D.

Lemma 1.3.3 ([49]): If the homogeneous boundary value problem

$$L[u] = 0 \text{ in } D$$

$$u = 0 \text{ on } \partial D \qquad (1.3.31)$$

$$u \in C^o(\bar{D}) \cap C^2(D)$$

has only the trivial solution $u \equiv 0$ in D then there exists a unique Hölder continuous solution $\mu(t)$ of (1.3.23).

Proof: We must show that the only solution of (1.3.30) is the trivial solution $\mu(t) = 0$ on ∂D. Let $\mu_o(t)$ be a solution of (1.3.30). Then

$$u_o(x,y) = \text{Re}\left[H_o(z)\phi_o(z) + \int_0^z H(z,t)\phi_o(t)dt\right] \qquad (1.3.32)$$

with

$$\phi_o(z) = \int_{\partial} \frac{t\mu_o(t)ds}{t-z}, \quad t \in \partial D \qquad (1.3.33)$$

is a solution of (1.3.31) which is continuous in $\bar{D}$ and satisfies $u_o = 0$ on ∂D. Hence from the hypothesis of the theorem $u_o(x,y) \equiv 0$ in D and hence $\phi_o(z) \equiv 0$ in D. But this is the case if and only if $\mu_o(t) = 0$ for t on ∂D (c.f. [42]) and the lemma is proved.

From the maximum principle the hypothesis of lemma 1.3.2 are satisfied if $c(x,y) \leqslant 0$ in D. The following lemma gives an alternative sufficient criteria for these hypothesis to be valid.

Lemma 1.3.4 ([49]): If there exists a real valued solution $v(x,y) \in C^o(\bar{D}) \cap C^2(D)$ of $L[u] = 0$ in D such that $v(x,y) \neq 0$ for $(x,y) \in \bar{D}$ then the homogeneous boundary value problem (1.3.31) has only the trivial solution.

Proof: Let $u(x,y)$ be a solution of (1.3.31). Then $w(x,y) = \dfrac{u(x,y)}{v(x,y)}$

is a solution of

$$w_{xx} + w_{yy} + (a+2\frac{\partial \log v}{\partial x})\frac{\partial w}{\partial x} + (b+2\frac{\partial \log v}{\partial y})\frac{\partial w}{\partial y} = 0 \qquad (1.3.34)$$

in D and $w(x,y) \in C^o(\bar{D}) \cap C^2(D)$, $w = 0$ on ∂D. Hence from the maximum principle $w(x,y) \equiv u(x,y) \equiv 0$ in D and the lemma is proved.

Definition 1.3.1: Let $v(x,y) = R(z_o,\bar{z}_o ; z,\bar{z})$ where $z_o = x_o+iy_o \in D$. Then $v(x_o,y_o) = 1$ and hence there exists a neighbourhood of (x_o,y_o) such that $v(x,y) > 0$ in this neighbourhood; such a neighbourhood will be called a Riemann neighbourhood of the point z_o.

Theorem 1.3.2 ([49]): Let $u(x,y)$ be a real valued solution of $L[u] = 0$ that is Hölder continuous in $\bar{D}$. Then $U(z,0)$ is Hölder continuous in $\bar{D}$.

Proof: Let $\gamma \subset \partial D$ be a closed arc such that γ lies entirely inside some Riemann neighbourhood. Complete this arc to form a closed curve $\partial D'$ such that D' is of class Ah, D' lies inside D, and $\bar{D}'$ lies in a Riemann neighbourhood. From lemma 1.3.4 and our previous analysis we can represent $u(x,y)$ inside D' as

$$u(x,y) = \int_{\partial D} \mu'(t)K(z,t)ds \qquad (1.3.35)$$

where $\mu'(t)$ is Hölder continuous on $\partial D'$. But from the representation (1.3.13) we see that

$$\phi(z) = \int_{\partial D'} \frac{t\mu'(t)ds}{t-z}, \quad z \in D' \quad . \qquad (1.3.36)$$

From the properties of Cauchy integrals we have that $\phi(z)$ is Hölder continuous on γ and hence so is $U(z,0)$. Covering ∂D by a finite number of overlapping closed arcs γ_i (each γ_i being contained in a Riemann neighbourhood) shows that $U(z,0)$ is Hölder continuous on ∂D and hence, from the properties of Cauchy integrals, Hölder continuous in D.

From Walsh's generalization of Runge's theorem (c.f. [50]) we have that $U(z,0)$ can be uniformly approximated in $\bar{D}$ by polynomials and we therefore have the following extension of Theorem 1.3.1:

Theorem 1.3.3 ([49]): Let $D \varepsilon Ah$ and let u_n be defined by (1.3.1). Then the set $\{u_n\}$ is complete in the maximum norm over D for the class of real valued solutions of $L[u] = 0$ which are Hölder continuous in $\bar{D}$. In order to apply Theorem 1.3.3 we need criteria for which a solution $u(x,y)$ of $L[u] = 0$ is Hölder continuous in $\bar{D}$. The following give criteria suitable for the purposes of these lectures.

Theorem 1.3.4 ([49]): Suppose $c(x,y) \leq 0$ in D, $D \varepsilon Ah$, and $u(x,y) \varepsilon C^o(\bar{D}) \cap C^2(D)$ is a real valued solution of $L[u] = 0$ in D such that

$$u(t) = f(t) \qquad \text{on } \partial D$$

where $f(t)$ is Hölder continuous on ∂D. Then $u(x,y)$ is Hölder continuous in $\bar{D}$.

Proof: This follows from the maximum principle, lemma 1.3.3, and (1.3.13), (1.3.18), since if $\mu(t)$ is Hölder continuous on ∂D then $\phi(z)$ is Hölder continuous in $\bar{D}$.

Theorem 1.3.5: Suppose ∂D has Hölder continuous curvature. Then if $u(x,y) \varepsilon C^o(\bar{D}) \cap C^2(D)$ is a real valued solution of $L[u] = 0$ in D such that

$$u(t) = 0 \qquad \text{on } \partial D$$

then $u(x,y)$ and its first and second derivative are Hölder continuous in $\bar{D}$.

Proof: This follows immediately from the Shauder estimates (c.f. [21]).

We state now the following generalization of Theorems 1.3.2 and 1.3.4, the proof of which can be found in [49].

Theorem 1.3.6 ([49]): Let $D \varepsilon Ah$ and $u(x,y) \varepsilon C^o(\bar{D}) \cap C^2(D)$ be a real valued solution of $L[u] = 0$ in D where $c(x,y) \leq 0$ in D and

$$u(t) = f(t) \qquad \text{on } \partial D,$$

where $\frac{df}{ds}$ is Hölder continuous on ∂D. Then $U(z,0)$ has a Hölder continuous first derivative in $\bar{D}$ (and hence $u(x,y)$ has Hölder continuous first derivatives in $\bar{D}$).

Further generalizations can be found in [47].

From Theorems 1.3.3 and 1.3.4 we can now approximate solutions to the Dirichlet problem for $L[u] = 0$ in D, where $D \epsilon Ah$, $c(x,y) \leqslant 0$, in the following manner : Orthonormalize the set $\{u_n\}$ in the L^2 norm over ∂D to obtain the complete set $\{\phi_n\}$ and set

$$c_n = \int_{\partial D} f\phi_n \tag{1.3.37}$$

$$u^N(x,t) = \sum_{n=0}^{N} c_n\phi_n \quad . \tag{1.3.38}$$

(Since Hölder continuous functions can be approximated by continuous functions we can, by the maximum principle, assume that $f(t)$ is merely continuous on ∂D and still conclude from Theorem 1.3.4 and Theorem 1.3.3 that the set $\{\phi_n\}$ is complete in the maximum norm over $\bar{D}$, and hence complete in the L^2 norm over $\bar{D}$). As we have already discussed in the introduction, we can now conclude that the given $\epsilon > 0$, N sufficiently large and D_o a compact subset of D,

$$\max_{D_o} |u-u^N| < \epsilon \ . \tag{1.3.39}$$

Since each ϕ_n is a solution of $L[u] = 0$ in D, error estimates can be found in the case when $c(x,y) \leqslant 0$ in D by applying the maximum principle.

In the next section we will discuss an alternate method for approximating solutions to $L[u] = 0$ by means of a complete family of solutions. The method to be discussed is based on the <u>Bergman kernel function</u> (c.f. [6]).

1.4 The Bergman Kernel Function.

We will restrict ourselves to the self-adjoint elliptic equation

$$u_{xx} + u_{yy} + q(x,y)u = 0 \qquad (1.4.1)$$

where $q(x,y)$ is an entire function of its independent complex variables. We again consider solutions $u(x,y)$ of (1.4.1) defined in a domain which is bounded, simply connected, and in class Ah, and make the assumption that $q(x,y) < 0$ in $\bar{D}$. Let $N(x,y\ ;\ \xi,\eta)$ and $G(x,y\ ;\ \xi,\eta)$ be the Neumann's and Green's function respectively of (1.4.1) in D. Then the <u>kernel function</u> $K(x,y\ ;\ \xi,\eta)$ of (1.4.1) in D is defined by

$$K(x,y\ ;\ \xi,\eta) = N(x,y\ ;\ \xi,\eta) - G(x,y\ ;\ \xi,\eta). \qquad (1.4.2)$$

Note that since the singularities of the singular parts of $N(x,y\ ;\ \xi,\eta)$ and $G(x,y\ ;\ \xi,\eta)$ cancel we have that $K(x,y\ ;\ \xi,\eta)$ is regular in D both as a function of (x,y) and (ξ,η). Furthermore, due to the symmetry of the Neumann's and Green's function, we have

$$K(x,y\ ;\ \xi,\eta) = K(\xi,\eta\ ;\ x,y). \qquad (1.4.3)$$

From the boundary conditions satisfied by the Neumann's and Green's function we have

$$\begin{aligned} K(x,y\ ;\ \xi,\eta) &= N(x,y\ ;\ \xi,\eta) \quad ; \quad (x,y)\in\partial D \\ \frac{\partial K}{\partial\nu}(x,y\ ;\ \xi,\eta) &= -\frac{\partial G}{\partial\nu}(x,y\ ;\ \xi,\eta) \quad ; \quad (x,y)\in\partial D \end{aligned} \qquad (1.4.4)$$

where ν is the unit inward normal to ∂D. Hence if $u(x,y)\in C^1(D)$ is a solution of (1.4.1) we have from Green's formulas

$$\begin{aligned} u(\xi,\eta) &= -\int_{\partial D} u(t)\,\frac{\partial K(t;\xi,\eta)}{\partial\nu}\,ds \\ u(\xi,\eta) &= -\int_{\partial D} K(t;\xi,\eta)\,\frac{\partial u(t)}{\partial\nu}\,ds \end{aligned} \qquad (1.4.5)$$

where $u(t) = u(x,y)$, $K(t;\xi,\eta) = K(x,y\ ;\ \xi,\eta)$ for $(x,y) = (x(t),y(t))\in\partial D$ and ds denotes arclength.

If we define the inner product $(\cdot,\cdot)_D$ by

$$(u,v)_D = \iint_D \left[u_x v_x + u_y v_y - quv\right] dxdy \tag{1.4.6}$$

where $u, c \varepsilon C^1(\bar{D})$ we see that $(\cdot,\cdot)_D$ satisfies all the conditions of an inner product. In particular since $q(x,y) < 0$ for $(x,y)\varepsilon \bar{D}$ we have $||u||_D^2 = (u,u)_D = 0$ if and only if $u \equiv 0$ in D. From Green's formula we have the fact that if $v(x,y)$ is a solution of (1.4.1) then

$$(u,v)_D = -\int_{\partial D} u \frac{\partial v}{\partial \nu} ds \quad , \tag{1.4.7}$$

in particular if $u(x,y)$ is a solution of (1.4.1) then (1.4.5) can be written as the single relation

$$u(\xi,\eta) = (u(x,y), K(x,y\ ;\ \xi,\eta))_D \tag{1.4.8}$$

since both $u(x,y)$ and $K(x,y\ ;\ \xi,\eta)$ are solutions of (1.4.1). Equation (1.4.8) is known as the <u>reproducing property</u> of the kernel function. In particular from Schwarz's inequality we have

$$\begin{aligned} |u(\xi,\eta)|^2 &= |(u,K)_D|^2 \\ &< (u,u)_D (K,K)_D \\ &= K(\xi,\eta\ ;\ \xi,\eta)||u||^2_D . \end{aligned} \tag{1.4.9}$$

Now let $\{u_n\}$ be the family of solutions to (1.4.1) defined by $u_{2n} = \text{Re}\ \underset{\sim}{B}_2\ \{z^n\}$, $u_{2n+1} = \text{Im}\ \underset{\sim}{B}_2\ \{z^n\}$, and orthonormalize this set with respect to $(\cdot,\cdot)_D$ to obtain the set $\{\phi_n\}$. From Theorem 1.3.6 we have that the set $\{\phi_n\}$ is complete in the Dirichlet norm $||\cdot||_D$ over $\bar{D}$ with respect to the class of solutions to (1.1.1) that are Hölder continuously differentiable. In particular if $u(x,y)$ is a solution of (1.4.1) which is Hölder continuously differentiable on ∂D then for any $\varepsilon > 0$ there exists an integer N and constants $a_0,\ldots,a_N$ such that

$$||u(x,y) - \sum_{n=0}^{N} a_n\phi_n(x,y)||^2_D < \varepsilon. \tag{1.4.10}$$

In particular since the set $\{\phi_n\}$ is an orthonormal set, the optimum choice of the constants $a_0, \ldots, a_N$ is given by

$$\begin{aligned} a_n &= (u,\phi_n)_D \\ &= -\int_{\partial D} u \frac{\partial\phi_n}{\partial\nu} ds \quad . \end{aligned} \tag{1.4.11}$$

From (1.4.9) and (1.4.10) we have

$$|u(\xi,\eta) - \sum_{n=0}^{N} a_n\phi_n(\xi,\eta)|^2 < \varepsilon\, K(\xi,\eta\ ;\ \xi,\eta)\ , \tag{1.4.12}$$

and hence the series

$$u(x,y) = \sum_{n=0}^{\infty} a_n\phi_n(x,y) \tag{1.4.13}$$

$$a_n = (u,\phi_n)_D$$

converges uniformly to $u(x,y)$ in every closed subdomain D_o of D. In particular setting $u(x,y) = K(x,y\ ;\ \xi,\eta)$ we have from (1.4.8)

$$(\phi_n(x,y),\ K(x,y\ ;\ \xi,\eta))_D = \phi_n(\xi,\eta) \tag{1.4.14}$$

and hence for (x,y) and (ξ,η) on compact subsets of D we have the remarkable representation

$$K(x,y\ ;\ \xi,\eta) = \sum_{n=0}^{\infty} \phi_n(x,y)\phi_n(\xi,\eta\ ). \tag{1.4.15}$$

Note that the representation (1.4.15) is in fact independent of the particular orthonormal system $\{\phi_n\}$ we started out with.

Numerical methods based on the kernel function can be found in [5]. Other numerical methods for solving the Dirichlet problem for elliptic equations using the method of integral operators can be found in [1], [24], [33], [34] and [47]

1.5 Inverse Methods in Compressible Fluid Flow.

In this section we will be considering stationary, irrotational flow of a two dimensional compressible fluid, and will derive an inverse method for obtaining flows past an obstacle due to a dipole at infinity. The approach we are about to derive is due to S.Bergman and our presentation is based on the material in [6]. Another excellent survey of the present topic can be found in [41].

Let $\vec{q}$ be the velocity vector of the motion and $\phi(x,y)$ be the velocity potential, i.e.

$$\vec{q} = -\operatorname{grad}\phi = (u,v) \tag{1.5.1}$$

where $u = -\dfrac{\partial\phi}{\partial x}$, $v = -\dfrac{\partial\phi}{\partial y}$. Let $\rho(x,y)$ denote the density of the fluid where $\rho = \rho(q^2)$, $q^2 = (\dfrac{\partial\phi}{\partial x})^2 + (\dfrac{\partial\phi}{\partial y})^2$. Then from the equation of continuity $\operatorname{div}(\rho\vec{q}) = 0$ we can assert the existence of a stream function $\psi(x,y)$ such that

$$\begin{aligned} \rho\,\frac{\partial\phi}{\partial x} &= \frac{\partial\psi}{\partial y} \\ \rho\,\frac{\partial\phi}{\partial y} &= -\frac{\partial\psi}{\partial x}, \end{aligned} \tag{1.5.2}$$

and where $\psi(x,y)$ remains constant along each stream line.

We now introduce the new variables

$$\begin{aligned} q &= (u^2+v^2)^{\frac{1}{2}} \\ \theta &= \arctan\frac{v}{u} \end{aligned} \tag{1.5.3}$$

(The inverse mapping is given by (c.f. [6]).

$$x = -\int \left(\frac{1}{q}\,(\cos\theta\phi_q - \frac{\sin\theta}{\rho}\,\psi_q)dq + \frac{1}{q}\,(\cos\theta\phi_\theta - \frac{\sin\theta}{\rho}\psi_\theta\,)d\theta \right)$$

$$y = -\int \left(\frac{1}{q}\,(\sin\theta\phi_q + \frac{\cos\theta}{\rho}\,\psi_q)dq + \frac{1}{q}\,(\sin\theta\phi_\theta + \frac{\cos\theta}{\rho}\psi_\theta\,)d\theta \right))$$

which are polar coordinates in the hodograph plane (u,v). Under this change of variables the nonlinear system (1.5.2) becomes the linear system (c.f. [6]).

$$\phi_\theta = \frac{q}{\rho}\psi_q \qquad (1.5.4)$$

$$\phi_q = -\left(1-\frac{q^2}{c^2}\right)\frac{1}{\rho q}\psi_\theta$$

where $c^2 = c^2(q^2)$ is the square of the local velocity of sound in the medium. If we now make the assumption that the flow is subsonic i.e. $q^2 < c^2$, and transform (1.5.4) into the pseudo-logarithmic plane (λ,θ) by means of the change of variables

$$\lambda = \int^q \frac{1}{q}\left(1-\frac{q^2}{c^2}\right)^{\frac{1}{2}} dq \qquad (1.5.5)$$

$$\theta = \theta$$

we arrive at the system

$$\phi_\theta = \ell(\lambda)\psi_\lambda$$
$$\phi_\lambda = -\ell(\lambda)\psi_\theta \qquad (1.5.6)$$

where
$$\ell(\lambda) = \frac{1}{\rho}\left(1-\frac{q^2}{c^2}\right)^{\frac{1}{2}} \qquad (1.5.7)$$

is a known function depending on the physical nature of the fluid under consideration. Note that in the case of an incompresible fluid flow $(c=\infty)$ the system (1.5.6) reduces (after the introduction of an appropriate scaling factor) to the Cauchy-Riemann equations. Eliminating $\phi(\lambda,\theta)$ from (1.5.6) gives

$$\ell(\lambda)\left[\psi_{\lambda\lambda} + \psi_{\theta\theta}\right] + \ell'(\lambda)\psi_\lambda = 0 \qquad (1.5.8)$$

where $\ell'(\lambda) = \frac{d\ell}{d\lambda}$. Setting $\psi = \ell^{\frac{1}{2}}\psi$ now gives

$$\psi_{\lambda\lambda} + \psi_{\theta\theta} - L(\lambda)\psi = 0 \qquad (1.5.9)$$

where
$$L(\lambda) = \frac{\Delta(\ell^{\frac{1}{2}})}{\ell^{\frac{1}{2}}}, \quad \Delta = \frac{\partial^2}{\partial\lambda^2} + \frac{\partial^2}{\partial\theta^2} .$$

We will now construct an integral operator which maps analytic functions into the class of solutions of (1.5.9). We will make the assumption (valid in particular for the case of an adiabatic gas where the pressure p is given by $p = \text{constant}\ \rho^{\gamma}$ for some constant γ) that $L(\lambda)$ is an analytic function of λ for $\lambda < 0$ and has a dominant of the form

$$L(\lambda) << C(\varepsilon-\lambda)^{-2} \tag{1.5.10}$$

where $C > 0$, $\varepsilon < 0$ ((1.5.10) is interpreted in the sense that $\left|\frac{d^n L(\lambda)}{d\lambda^n}\right| \leqslant C \frac{d^n}{d\lambda^n} (\varepsilon-\lambda)^{-2}$ for $n = 0, 1,2, \ldots$.) Note that the integral operators previously constructed are not applicable in this special case, since $L(\lambda)$ is not in general an entire function of λ. We first look for a formal solution of (1.5.9) in the form

$$\psi(\lambda,\theta) = \sum_{n=0}^{\infty} G_n(\lambda) g_n(\lambda,\theta) \tag{1.5.11}$$

where $g_n(\lambda,\theta)$ is a harmonic function of λ and θ.

Proceeding formally we have

$$0 = \Delta\psi - L\psi = \sum_{n=0}^{\infty} (g_n(G''_n - LG_n) + 2 \frac{\partial g_n}{\partial\lambda} G'_n \tag{1.5.12}$$

and hence we require

$$G_0 = 1$$

$$G'_{n+1} = G''_n - LG_n \quad ; \quad n = 1,2, \ldots \tag{1.5.13}$$

$$2 \frac{\partial g_n}{\partial\lambda} = - g_{n-1} \tag{1.5.14}$$

with $g_0(\lambda,\theta)$ an arbitrary harmonic function. We normalize the $G_n(\lambda)$ by imposing the condition

$$G_n(-\infty) = 0, \quad n = 1,2, \ldots$$

which is motivated by the fact that $\lambda = -\infty$ corresponds to $q = 0$, the state of rest of the fluid.

We first solve the recursion relation (1.5.14). Let $\phi_o(\zeta)$ be an analytic function of $\zeta = \lambda+i\theta$ and let $g_o(\lambda,\theta) = \text{Re } \phi_o$. Then (1.5.14) will hold if

$$g_n(\lambda,\theta) = \text{Re } \Phi_n(\zeta) \tag{1.5.15}$$

where

$$\frac{d\Phi_n(\zeta)}{d\zeta} = -\frac{1}{2}\Phi_{n-1}(\zeta), \quad n = 1,2,\ldots \quad . \tag{1.5.16}$$

In particular a solution of (1.5.16) is given by

$$\Phi_n(\zeta) = \frac{(-1)^n}{(n-1)!2^n} \int_o^\zeta \Phi_o(t)(\zeta-t)^{n-1}dt \tag{1.5.17}$$

and hence

$$g_n(\lambda,\theta) = \frac{(-1)^n}{(n-1)!2^n} \text{Re } \left(\int_o^\zeta \Phi_o(t)(\zeta-t)^{n-1}dt \right) \tag{1.5.18}$$

for n = 1,2,... . A formal solution of (1.5.9) is thus given by

$$\psi(\lambda,\theta) = \text{Re}\left(\Phi_o(\zeta) + \int_o^\zeta \Phi_o(t)U(\lambda,\theta;t)dt\right) \tag{1.5.19}$$

where

$$U(\lambda,\theta;t) = \sum_{n=1}^{\infty} \frac{(-1)^n}{(n-1)!2^n} G_n(\lambda)(\zeta-t)^{n-1}. \tag{1.5.20}$$

Our formal analysis will now be valid provided we can show that the series (1.5.20) converges uniformly to an analytic function for λ and θ in the region of definition of $\psi(\lambda,\theta)$ and for t in the region of integration in (1.5.19). We will do this through the method of dominants. Define the functions $Q_n(\lambda)$ by the recursive scheme

$$\begin{aligned} &Q_o = 1 \\ &Q'_{n+1} = Q''_n + C(\varepsilon-\lambda)^{-2}Q_n \quad ; \quad n = 1,2, \ldots \\ &Q_n(-\infty) = 0 \ ; \ n = 1,2, \ \ldots \quad . \end{aligned} \tag{1.5.21}$$

The system (1.5.21) can be solved explicitly by setting

$$Q_n(\lambda) = n!\ (\varepsilon-\lambda)^{-n}\mu_n \tag{1.5.22}$$

where the μ_n satisfy the recursion formula

$$\mu_o = 1$$

$$\mu_{n+1} = \mu_n \frac{(n+\alpha)(n+\beta)}{(n+1)^2} \tag{1.5.23}$$

with

$$\alpha = \frac{1}{2} - (\frac{1}{4} - C)^{\frac{1}{2}}$$
$$\beta = \frac{1}{2} + (\frac{1}{4} - C)^{\frac{1}{2}} \quad . \tag{1.5.24}$$

From (1.5.10) and (1.5.13), (1.5.15) it is easily seen that

$$G_n(\lambda) << Q_n(\lambda) , \tag{1.5.25}$$

and hence the series (1.5.20) is majorized by the series

$$\Omega(\lambda,\theta;t) = \sum_{n=1}^{\infty} \frac{Q_n(\lambda)}{(n-1)!2^n} |\zeta-t|^{n-1}$$

$$= \sum_{n=1}^{\infty} n\ \mu_n \frac{|\zeta-t|^{n-1}}{2^n(\varepsilon-\lambda)^n}$$

$$= \frac{1}{2}(\varepsilon-\lambda)^{-1} \sum_{n=1}^{\infty} n\ \mu_n (\frac{|\zeta-t|}{2(\varepsilon-\lambda)})^{n-1} \tag{1.5.26}$$

$$= \frac{1}{2(\varepsilon-\lambda)} \frac{d}{dx} F(\alpha,\beta,1; \frac{|\zeta-t|}{2(\varepsilon-\lambda)})$$

where $F(\alpha,\beta,1;x)$ is the hypergeometric function of Gauss. But it is well known that the hypergeometric series for $F(\alpha,\beta,1;x)$ coverges uniformly for $|x| \leqslant a < 1$ provided α and β are not zero or a positive integer, which from (1.5.24) is certainly not the case here. Hence the series (1.5.26) (and hence the series (1.5.20)) converges uniformly for

$$\frac{|\zeta-t|}{2(\varepsilon-\lambda)} \leqslant a < 1 \quad , \lambda < \varepsilon < 0 \tag{1.5.27}$$

i.e. ζ must satisfy

$$|\zeta| \leq 2a|\lambda| < 2|\lambda| \tag{1.5.28}$$

and since $\zeta = \lambda+i\theta$ we have

$$|\theta| < \sqrt{3}|\lambda| \tag{1.5.29}$$

i.e. ζ must lie in an angle of 120° symmetric to the λ axis in the left half plane. Conversely, if ζ lies in this angular region it can always be connected with the origin by a path along which t fulfills (1.5.27) for an appropriate value of $a < 1$. Hence we can construct solutions of (1.5.9) in the region (1.5.29) by means of (1.5.19).

We will now show how the operator defined by (1.5.19) can serve as the basis for the development of an inverse approach to solving boundary value problems in the theory of subsonic, compressible fluid flow. We restrict our attention to the case in which the flow domain in the physical (x,y) plane contains the point at infinity and in which the flow originates from a dipole there (i.e. the velocity at infinite is uniform). Let the flow domain be bounded by a closed curve C, and let

$$F(z) = \phi+i\psi = Az+a_o + \frac{a_1}{z} + \ldots \tag{1.5.30}$$

be the complex velocity potential in the case of <u>incompressible</u> flow (expanded about the dipole at infinity). The function F(z) can be obtained by classical methods in analytic function theory (c.f. [6]). Note that on C we have $\psi = 0$. The velocity function of the incompressible flow near infinity is now given by

$$w = -F'(z) = u-iv = -A + \frac{a_1}{z^2} + \ldots, \tag{1.5.31}$$

and since in the case of an incompressible fluid flow $\lambda = \log q$ we have

$$\zeta = \lambda+i\theta = \log \bar{w} . \tag{1.5.32}$$

From (1.5.31) and (1.5.32) we have

$$\zeta = \log(-\bar{A}) - \left(\frac{\bar{a}_1}{\bar{A}} \right) \frac{1}{\bar{z}^2} + \ldots \tag{1.5.33}$$

and hence for ζ near $\bar{\beta} = \log(-\bar{A})$

$$z = \frac{a}{(\bar{\zeta}-\beta)^{\frac{1}{2}}} + P((\bar{\zeta}-\beta)^{\frac{1}{2}}) \tag{1.5.34}$$

where $P(\tau)$ is a power series in τ about the origin and $a = -\left(\frac{a_1}{A} \right)^{\frac{1}{2}}$.

In the pseudo-logarithmic plane we thus obtain a complex potential $F^*(\zeta)$ defined by means of

$$\overline{F^*(\zeta)} = F(z(\bar{\zeta})) = \frac{A^*}{(\bar{\zeta}-\beta)^{\frac{1}{2}}} + P^*((\bar{\zeta}-\beta)^{\frac{1}{2}}) \tag{1.5.35}$$

where $P^*(\tau)$ is a power series in τ about the origin and A^* is a constant. From (1.5.33) we see that the image of the flow domain in the (λ,θ) plane covers this plane in a non-schlicht manner and has the point $\zeta = \log(-\bar{A})$ as a second order branch point. The stream function $\psi(\lambda,\theta)$ is defined by

$$\psi(\lambda,\theta) = \mathrm{Re}(iF^*(\zeta)) \tag{1.5.36}$$

(c.f. (1.5.30) and (1.5.35))and vanishes on the boundary ∂D of the image of the flow domain D in the ζ plane.

Now associate with $iF^*(\zeta) = \Phi_0(\zeta)$ a solution $\Psi(\lambda,\theta)$ of (1.5.9) defined by (1.5.19) (we will assume that the image of the flow domain in the pseudo-logarithmic plane lies entirely in the region where the operator (1.5.19) is applicable). Note that for small velocities $\ell(\lambda) \approx$ constant and hence $L(\lambda)$ can be assumed to be small for large negative values of λ. This in turn implies the constant C in (1.5.10) is small and hence from the recursion relations (1.5.23), (1.5.24) $U(\lambda,\theta;t)$ is small for large negative values of λ, i.e. for such values of λ $\Psi(\lambda,\theta)$ does not differ too much from $\mathrm{Re}(\Phi_0(\zeta))$, the corresponding solution in the incompressible case. $\Psi(\lambda,\theta)$ has a singularity at the point $\log(-\bar{A}) = \zeta$ which corresponds to a dipole of the compressible fluid flow; however $\Psi(\lambda,\theta)$ will not in general be zero on the boundary ∂D of the non-schlicht domain D.

Having obtained a solution $\Psi(\lambda,\theta)$ of (1.5.9) with prescribed singularity at $\zeta = \log(-\bar{A})$, we now have to see what type of flow this solution represents in the physical (x,y) plane. It can be easily seen that the point $\zeta = \log(-\bar{A})$ in the (λ,θ) plane corresponds to the point at infinity in the (x,y) plane and that the flow behaves there as if a dipole were situated at infinity. The curve C* in the (x,y) plane on which ψ vanishes will of course be different in general from the original curve C for which we wanted to solve the boundary value problem; however C* will not be too different from C if the velocities involved are not too near the sonic velocity $q = c$. By choosing complex velocity potentials associated with different curves C and constructing $\Psi(\lambda,\theta)$ as above we now have an inverse method for obtaining subsonic compressible flows past an obstacle originating from a dipole at infinity.

For an example of numerical experiments using the methods of this section see [3] and [4] .

The basic ideas of the approach described above for subsonic fluid flow can also be used to study transonic flow problems. Particular problems of course arise due to the need to continue the solution past the sonic line, and the analysis is by no means trivial. However these difficulties have been overcome and the use of integral operators and inverse methods has recently led to the numerical design of shock-free transonic flow at specified cruising speeds. The interested reader is referred to

S. Bergman, Two-dimensional transonic flow patterns, Amer.J.Math.70(1948),856-891.

P.R. Garabedian and D.G. Korn, Numerical design of transonic airfoils, in Numerical Solution of Partial Differential Equations-II, Bert Hubbard, editor, Academic Press, 1971, 253-271.

D.G. Korn, Transonic design in two dimensions, in Constructive and Computational Methods for Differential and Integral Equations, D.L. Colton and R.P. Gilbert, editors, Springer-Verlag Lecture Note Series Vol.430, Springer-Verlag, 1974, 271-288.

II Parabolic equations in one space variable

2.1 Integral Operators.

We now want to develop a theory for parabolic equations in one space variables that is analegous to the theory just developed for elliptic equations in two independent variables. To this end we consider the general linear homogeneous parabolic equation of the second order in one space variable written in normal form as

$$u_{xx} + a(x,t)u_x + b(x,t)u - c(x,t)u_t = 0 \quad . \tag{2.1.1}$$

In the theory we are about to develop we will need to construct a variety of integral operators for (2.1.1), and in each such construction we will impose somewhat different assumptions on the coefficients of (2.1.1). The first operator we will consider will map ordered pairs of analytic functions of a single complex variable onto analytic solutions of (2.1.1). In order to construct this operator we will make the assumption that the coefficients $a(x,t)$, $b(x,t)$ and $c(x,t)$ in (2.1.1) are analytic functions of the complex variables x and t for $|x| < \infty$ and $|t| < t_o$ where t_o is a positive constant. By making the change of dependent variable

$$u(x,t) = v(x,t) \exp \left\{ -\frac{1}{2} \int_0^x a(\xi,t)d\xi \right\} \tag{2.1.2}$$

we arrive at an equation for $v(x,t)$ of the same form as (2.1.1) but with $a(x,t)=0$. Hence without loss of generality we can restrict our attention to equations of the form

$$L[u] \equiv u_{xx} + b(x,t)u - c(x,t)u_t = 0 \tag{2.1.3}$$

where $b(x,t)$ and $c(x,t)$ are analytic functions of x and t for $|x| < \infty$, $|t| < t_o$.

We now look for a solution of (2.1.3) in the form

$$u(x,t) = -\frac{1}{2\pi i}\oint_{|t-\tau|=\delta} E^{(1)}(x,t,\tau)f(\tau)d\tau - \frac{1}{2\pi i}\oint_{|t-\tau|=\delta} E^{(2)}(x,t,\tau)g(\tau)d\tau \tag{2.1.4}$$

where $t_o-|t| > \delta > 0$ and $f(\tau)$ and $g(\tau)$ are arbitrary analytic functions of τ for $|\tau| < t_o$. We will furthermore ask that $E^{(1)}(x,t,\tau)$ and $E^{(2)}(x,t,\tau)$ satisfy the initial conditions

$$E^{(1)}(0,t,\tau) = \frac{1}{t-\tau} \tag{2.1.5a}$$

$$E_x^{(1)}(0,t,\tau) = 0 \tag{2.1.5b}$$

$$E^{(2)}(0,t,\tau) = 0 \tag{2.1.6a}$$

$$E_x^{(2)}(0,t,\tau) = \frac{1}{t-\tau} \tag{2.1.6b}$$

and be analytic functions of their independent variables for $|x| < \infty$, $|t| < t_o$, $|\tau| < t_o$, $t \neq \tau$. We shall first construct the function $E^{(1)}(x,t,\tau)$. Setting $g(\tau)=0$ and substituting (2.1.4) into $L[u]=0$ shows that, as a function of x and t, $E^{(1)}(x,t,\tau)$ must be a solution of $L[u]=0$ for $t \neq \tau$. We now assume that $E^{(1)}(x,t,\tau)$ has the expansion

$$E^{(1)}(x,t,\tau) = \frac{1}{t-\tau} + \sum_{n=2}^{\infty} x^n p^{(n)}(x,t,\tau) \tag{2.1.7}$$

where the $p^{(n)}(x,t,\tau)$ are analytic functions to be determined. Note that if termwise differentiation is permitted the series (2.1.7) satisfies the initial conditions (2.1.5a) and (2.1.5b). Substituting (2.1.7) into (2.1.3) we are led to the following recursion formula for the $p^{(n)}(x,t,\tau)$:

$$p^{(1)} = 0$$

$$p^{(2)} = -\frac{c}{2(t-\tau)^2} - \frac{b}{2(t-\tau)} \tag{2.1.8}$$

$$p^{(k+2)} = -\frac{2}{k+2}\, p_x^{(k+1)} - \frac{1}{(k+2)(k+1)}\left[p_{xx}^{(k)} + bp^{(k)} - cp_t^{(k)}\right] \; ; \; k \geq 1 \; .$$

If we now define $Q^{(k)}(x,t,\tau)$ by the equation

$$Q^{(k)}(x,t,\tau) = \tau^k p^{(k)}(x,t,t-\tau) \tag{2.1.9}$$

then (2.1.8) yields the following recursion formula for the $Q^{(k)}(x,t,\tau)$:

$$Q^{(1)} = 0$$

$$Q^{(2)} = -\frac{1}{2}\left[c+\tau b\right]$$

$$Q^{(k+2)} = -\frac{2\tau}{k+2}Q_x^{(k+1)} - \frac{2\tau}{(k+2)(k+1)}\Big[\tau Q_{xx}^{(k)} + \tau b\,Q^{(k)} - \tau c\,Q_t^{(k)} + ck\,Q^{(k)} - \tau c\,Q_\tau^{(k)}\Big] \; ; \; k \geq 1 \; . \tag{2.1.10}$$

Now let M_o be a positive constant such that

$$c(x,t) \ll M_o(1-\frac{x}{r})^{-1}(1-\frac{t}{t_o})^{-1}$$

$$b(x,t) \ll M_o(1-\frac{x}{r})^{-1}(1-\frac{t}{t_o})^{-1} \tag{2.1.11}$$

for $|x| < r$ and $|t| < t_o$. Using the fact that

$$\tau \ll 2\,t_o(1-\frac{t}{2t_o})^{-1} \tag{2.1.12}$$

we shall now show by induction that there exist positive constants M_n, $n=1,2,\ldots$, and ε (where ε can be chosen arbitrarily small and is independent of n and M_n is a bounded function of n) such that for $|x| < r$, $|t| < t_o$, $|\tau| < 2t_o$, we have

$$Q^{(n+1)} \ll M_{n+1}\,4^{n+1}\,t_o^{n+1}\,(\frac{3}{2}+\varepsilon)^{n+1} \cdot (1-\frac{x}{r})^{-(n+1)}(1-\frac{t}{t_o})^{-(n+1)}\,(1-\frac{\tau}{2t_o})^{-(2n+2)}r^{-(n+1)} \, , \tag{2.1.13}$$

$n=0,1,2,\ldots$. (2.1.13) is clearly true for $n=0$ and $n=1$, and from (2.1.10) it can be shown that (2.1.13) is true for $n=k+1$ if the M_n are defined by the recursion formula

$$M_{n+2} = (\frac{3}{2} + \varepsilon)^{-1} \left[M_{n+1} + \frac{M_n}{(\frac{3}{2} + \varepsilon)} \left(\frac{n}{2(n+2)} + \frac{M_o r^2}{2(n+2)(n+1)} + \frac{nM_o r^2}{(n+2)(n+1)t_o}\right)\right] \tag{2.1.14}$$

The proof of (2.1.13) now follows by induction once we have shown that M_n is a bounded function of n. For $n \geqslant n_o = n_o(\varepsilon)$ we have from (2.1.14) that

$$M_{n+2} \leqslant (\frac{3}{2} + \varepsilon)^{-1} \left[M_{n+1} + \frac{M_n}{(\frac{3}{2} + \varepsilon)} (\frac{1}{2} + \frac{\varepsilon}{2})\right] ; \; n \geqslant n_o . \tag{2.1.15}$$

If $M_{n+1} \leqslant M_n$ for $n \geqslant n_o$ we are done, for then we have $M_n \leqslant \max \{ M_1, M_2, \ldots, M_{n_o} \}$. Suppose then that there exists $n_1 \geqslant n_o$ such that $M_{n_1+1} > M_{n_1}$. Then from (2.1.15) we have

$$\begin{aligned} M_{n_1+2} &< (\frac{3}{2} + \varepsilon)^{-1} \left[M_{n_1+1} + M_{n_1+1} \frac{(\frac{1}{2} + \frac{\varepsilon}{2})}{(\frac{3}{2} + \varepsilon)} \right] \\ &= \frac{(2 + \frac{3\varepsilon}{2})}{(\frac{3}{2} + \varepsilon)(\frac{3}{2} + \varepsilon)} M_{n_1+1} \qquad (2.1.16) \\ &< M_{n_1+1} \end{aligned}$$

and by induction

$$M_{n_1+m} \leqslant M_{n_1+1} \tag{2.1.17}$$

for m=1,2,3,... . Hence $M_n \leqslant \max \{ M_1, M_2, \ldots M_{n_1+1}\}$ and we can conclude that M_n is a bounded function of n.

We now return to the convergence of the series (2.1.7). Let δ_o, δ_1, and $\alpha > 1$ be positive numbers and let

$$\begin{aligned} &|x| \leqslant r/\alpha && |\tau| \leqslant t_o \\ &|t| \leqslant t_o/(1+\delta_1) && |t-\tau| \geqslant \delta_o \end{aligned} \tag{2.1.18}$$

Then

$$(1-\frac{x}{r}) \geqslant \frac{\alpha-1}{\alpha} \qquad (1-\frac{\tau}{2t_o}) \geqslant \frac{1}{2} \qquad (2.1.19)$$

$$(1-\frac{t}{t_o}) \geqslant \frac{\delta_1}{(1+\delta_1)} \qquad |t-\tau| < 2t_o$$

Hence for x,t,τ restricted as in (2.1.18) we have from (2.1.13) that the series (2.1.7) is majorized by

$$\frac{1}{\delta_o} + \sum_{n=2}^{\infty} \frac{M_n 16^n t_o^{\,n}(\frac{3}{2}+\varepsilon)^n(\alpha-1)^n(1+\delta_1)^n}{\alpha^{2n}\delta_o^n\delta_1^n} \qquad (2.1.20)$$

Due to the fact that M_n is a bounded function of n, it is seen that if α is chosen sufficiently large then the series (2.1.20) converges. Since δ_o,δ_1 and ε are arbitrarily small (and independent of r) and r can be chosen arbitrarily large, we can now conclude that the series (2.1.20) converges uniformly and absolutely for $|x| \leqslant r$, $|t| \leqslant t_o/(1+\delta_1)$, $|\tau| \leqslant t_o$ and $|t-\tau| \geqslant \delta_o$ for δ_o and δ_1 arbitrarily small and r arbitrarily large. Since each term of the series (2.1.20) is an analytic function of x, t and τ for $|x| < \infty$, $|t| < t_o$, $|\tau| < t_o$, $\tau \neq t$, we can conclude that $E^{(1)}(x,t,\tau)$ exists and is an analytic function of its independent variables for $|x| < \infty$, $|t| < t_o$, $|\tau| < t_o$ and $t \neq \tau$. At the point $t=\tau$, $E^{(1)}(x,t,\tau)$ has an essential singularity. It is clear from the above discussion that termwise differentiation of the series (2.1.7) is permissible and hence $E^{(1)}(x,t,\tau)$ satisfies the differential equation (2.1.3) and the initial conditions (2.1.5a) and (2.1.5b).

We now turn our attention to the construction of the function $E^{(2)}(x,t,\tau)$. Setting $f(\tau)=0$ in (2.1.4) and substituting this equation into (2.1.3) shows that, as a function of x and t, $E^{(2)}(x,t,\tau)$ must be a solution of $L[u]=0$ for $t \neq \tau$. We now assume that $E^{(2)}(x,t,\tau)$ has the expansion

$$E^{(2)}(x,t,\tau) = \frac{x}{t-\tau} + \sum_{n=3}^{\infty} x^n q^{(n)}(x,t,\tau) \tag{2.1.21}$$

where the $q^{(n)}(x,t,\tau)$ are analytic functions (except for $t=\tau$) to be determined. We again note that if termwise differentiation is permitted the series (2.1.21) satisfies the initial conditions (2.1.6a), (2.1.6b). Substituting (2.1.21) into (2.1.3) leads to the following recursion formulas for the $q^{(n)}(x,t,\tau)$:

$$\begin{aligned} q^{(2)} &= 0 \\ q^{(3)} &= -\frac{c}{6(t-\tau)^2} - \frac{b}{6(t-\tau)} \end{aligned} \tag{2.1.22}$$

$$q^{(k+2)} = -\frac{2}{k+2} q_x^{(k+1)} - \frac{1}{(k+2)(k+1)} \left[q_{xx}^{(k)} + bq^{(k)} - cq_t^{(k)}\right] \; ; \; k \geq 2.$$

The recursion scheme (2.1.22) is almost identical to the scheme given in (2.1. 8), and following our previous analysis we can again verify that the series (2.1.21) defines an analytic function of x,t and τ for $|x| < \infty$, $|t| < t_0$, $|\tau| < t_0$, $t \neq \tau$, which satisfies $L[u]=0$ for $t \neq \tau$ and the initial data (2.1.6a), (2.1.6b). At the point $t=\tau$, $E^{(2)}(x,t,\tau)$ has an essential singularity. It is of interest to contrast this singular nature of the functions $E^{(1)}(x,t,\tau)$ and $E^{(2)}(x,t,\tau)$ with the analytic nature of the generating function of the Bergman operator $\underset{\sim}{B}_2$ for elliptic equations.

We have now shown that the integral operator defined by

$$\begin{aligned} u(x,t) = \underset{\sim}{P}_1\{f,g\} &= -\frac{1}{2\pi i} \oint_{|t-\tau|=\delta} E^{(1)}(x,t,\tau) f(\tau) d\tau \\ &\quad - \frac{1}{2\pi i} \oint_{|t-\tau|=\delta} E^{(2)}(x,t,\tau) g(\tau) d\tau \end{aligned} \tag{2.1.23}$$

exists and maps ordered pairs of analytic functions into the class of analytic solutions of $L[u]=0$. It is a simple matter to show that in fact

<u>every</u> solution of $L[u]=0$ which is analytic for $|t| < t_0$, $|x| < x_0$, where t_0 and x_0 are positive constants, can be represented in the form (2.1.23). For let $u(x,t)$ be an analytic solution of $L[u]=0$ for $|x| < x_0$, $|t| < t_0$, and set $u(0,\tau)=f(\tau)$, $u_x(0,\tau)=g(\tau)$. Then $f(\tau)$ and $g(\tau)$ are analytic for $|\tau| < t_0$. Define $w(x,t)$ by

$$w(x,t) = \underset{\sim}{P}_1\{f,g\} . \tag{2.1.24}$$

Then $w(x,t)$ is an analytic solution of $L[u]=0$ and from (2.1.5a), (2.1.5b), (2.1.6a), (2.1.6b) we have

$$w(0,t) = -\frac{1}{2\pi i}\oint_{|t-\tau|=\delta} \frac{f(\tau)}{t-\tau}\, d\tau = f(t) \tag{2.1.25}$$

$$w_x(0,t) = -\frac{1}{2\pi i}\oint_{|t-\tau|=\delta} \frac{g(\tau)}{t-\tau}\, d\tau = g(t) \quad ;$$

i.e. the Cauchy data for $w(x,t)$ and $u(x,t)$ argree on the noncharacteristic curve $x=0$. From the Cauchy-Kowalewski theorem (c.f. [21]) we can now conclude that $u(x,t) = w(x,t)$, i.e. $u(x,t)$ can be represented in the form (2.1.23). We summarize our results in the following theorem:

<u>Theorem 2.1.1.</u> ([9]): Let the coefficients $b(x,t)$ and $c(x,t)$ of (2.1.3) be analytic functions of the complex variables x and t for $|x| < \infty$, $|t| < t_0$. Then if $u(x,t)$ is a solution of (2.1.3) which is analytic for $|x| < x_0$, $|t| < t_0$, $u(x,t)$ can be prepresented in the form $u(x,t)=\underset{\sim}{P}_1\{f,g\}$ where $f(t)=u(0,t)$ and $g(t)=u_x(0,t)$ are analytic functions of t for $|t| < t_0$. Conversely if $f(t)$ and $g(t)$ are analytic for $|t| < t_0$ then $u(x,t) = \underset{\sim}{P}_1\{f,g\}$ is a solution of (2.1.3) which is an analytic function of x and t for $|x| < \infty$, $|t| < t_0$.

The integral operator $\underset{\sim}{P}_1$ suffers from the disadvantage that its range is the class of analytic solutions of $L[u]=0$. However it is known from the general theory of parabolic equations that solutions of $L[u]=0$ are in general not analytic in the t variable, even though the coefficients of $L[u]=0$ are analytic functions of x and t (c.f. [26]).

Hence we will now construct a class of operators whose range is the class of solutions to (2.1.1) which are twice continuously differentiable with respect to x and continuously differentiable with respect to t. Such solutions will be called classical. We assume that in (2.1.1) we have $c(x,t) > 0$, and make the change of variables

$$\xi = \int_0^x \sqrt{c(s,t)}\,ds \qquad (2.1.26)$$

$$\tau = t$$

This transformation reduces (2.1.1) to an equation of the same form but with $c(x,t)=1$. If we now make a change of variables of the form (2.1.2) we arrive at an equation of the form

$$u_{xx} + q(x,t)u = u_t \qquad (2.1.27)$$

and we will henceforth restrict our attention to parabolic equations which are written in this canonical form. We will first consider classical solutions $u(x,t)$ of (2.1.27) defined in the rectangle $R^+ = \{(x,t) : 0 < x < x_o, 0 < t < t_o\}$ such that $u(x,t)$ is continuously differentiable for $0 \leq x < x_o$, $0 < t < t_o$, and make the assumptions that $q(x,t)$ is continuously differentiable for $-x_o < x < x_o$, $|t| < t_o$, and is an analytic function of t for $|t| < t_o$. Here x_o and t_o are again positive constants.

Remark 1: The assumptions on $q(x,t)$ can be weakened.

Remark 2: The operators we are about to consider are related to the translation operators of Levitan for ordinary differential equations (c.f. [40]

We now look for solutions of (2.1.27) in the form

$$u(x,t) = h(x,t) + \int_0^x K(s,x,t)h(s,t)ds \tag{2.1.28}$$

where h(x,t) is a classical solution of the heat equation

$$h_{xx} = h_t \tag{2.1.29}$$

in R^+, is continuously differential for $0 \leq x < x_0$, $0 < t < t_0$, and satisfies the Dirichlet data $h(0,t)=0$ (Note that from (2.1.28) this implies that $u(0,t)=0$ also). Substituting (2.1.28) into (2.1.27) and integrating by parts shows that (2.1.28) is a solution of (2.1.27) provided K(s,x,t) is a solution of

$$K_{xx} - K_{ss} + q(x,t)K = K_t \tag{2.1.30}$$

for $0 < s \leq x < x_0$ which satisfies the initial data

$$K(x,x,t) = -\frac{1}{2}\int_0^x q(s,t)ds \tag{2.1.31a}$$

$$K(0,x,t) = 0 \quad . \tag{2.1.31b}$$

Now suppose that instead of satisfying $h(0,t)=0$, h(x,t) satisfies $h_x(0,t)=0$. We again look for a solution of (2.1. 27) in the form

$$u(x,t) = h(x,t) + \int_0^x M(s,x,t)h(s,t)ds \quad . \tag{2.1.32}$$

Then it is seen that (2.1.32) will be a solution of (2.1.27) provided M(s,x,t) is a solution of

$$M_{xx} - M_{ss} + q(x,t)M = M_t \tag{2.1.33}$$

for $0 < s \leq x < x_0$ which satisfies the initial data

$$M(x,x,t) = -\frac{1}{2}\int_0^x q(s,t)ds \tag{2.1.34a}$$

$$M_s(0,x,t) = 0 \quad . \tag{2.1.34b}$$

If the functions K(s,x,t) and M(s,x,t) exist, we can now define two operators $\underset{\sim}{T}_1$ and $\underset{\sim}{T}_2$ mapping solutions of the heat equation onto solutions of

(2.1.27) by

$$\underset{\sim}{T}_1\{h\} = h(x,t) + \int_0^x K(s,x,t)h(s,t)ds \tag{2.1.35}$$

$$\underset{\sim}{T}_2\{h\} = h(x,t) + \int_0^x M(s,x,t)h(s,t)ds \tag{2.1.36}$$

where the domain of $\underset{\sim}{T}_1$ is the class of solutions to the heat equation in R^+ satisfying $h(0,t)=0$ and the domain of $\underset{\sim}{T}_2$ is the class of solutions to the heat equation in R^+ satisfying $h_x(0,t)=0$.

We will now show the existence of the functions $K(s,x,t)$ and $M(s,x,t)$. Due to the regularity assumptions on $q(x,t)$ we will in fact show that $K(s,x,t)$ and $M(s,x,t)$ are twice continuously differentiable solutions of (2.1.30) and (2.1.33) for $-x_o < x < x_o$, $-x_o < s < x_o$, $|t| < t_o$. Suppose $E(s,x,t)$ satisfies

$$E_{xx} - E_{ss} + q(x,t)E = E_t \tag{2.1.37}$$

for $-x_o < x < x_o$, $-x_o < s < x_o$, $|t| < t_o$, and assumes the initial data

$$E(x,x,t) = -\frac{1}{2}\int_0^x q(s,t)ds \tag{2.1.38a}$$

$$E(-x,x,t) = \frac{1}{2}\int_0^x q(s,t)ds \tag{2.1.38b}$$

Then

$$K(s,x,t) = \frac{1}{2}\left[E(s,x,t) - E(-s,x,t)\right] \tag{2.1.39}$$

satisfies (2.1.30) and (2.1.31a), (2.1.31b). Similarly, if $G(s,x,t)$ satisfie

$$G_{xx} - G_{ss} + q(x,t)G = G_t \tag{2.1.40}$$

for $-x_o < x < x_o$, $-x_o < s < x_o$, $|t| < t_o$, and assumes the initial data

$$G(x,x,t) = -\frac{1}{2}\int_0^x q(s,t)ds \tag{2.1.41a}$$

$$G(-x,x,t) = -\frac{1}{2}\int_0^x q(s,t)ds \tag{2.1.41b}$$

then

$$M(s,x,t) = \frac{1}{2}\left[G(s,x,t) + G(-s,x,t)\right] \tag{2.1.42}$$

satisfies (2.1.33) and (2.1.34a), (2.1.34b). Hence it suffices to show the existence of the functions E(s,x,t) and G(s,x,t). We will now do this for E(s,x,t); the existence of G(s,x,t) follows in an identical fashion. Let

$$x = \xi+\eta \tag{2.1.43}$$
$$s = \xi-\eta$$

and define $\tilde{E}(\xi,\eta,t)$ and $\tilde{q}(\xi,\eta,t)$ by

$$\begin{aligned}\tilde{E}(\xi,\eta,t) &= E(\xi-\eta,\ \xi+\eta,t)\\ \tilde{q}(\xi,\eta,t) &= q(\xi+\eta,t).\end{aligned} \tag{2.1.44}$$

Then (2.1.37), (2.1.38a), (2.1.38b) become

$$\tilde{E}_{\xi\eta} + \tilde{q}(\xi,\eta,t)\tilde{E} = \tilde{E}_t \tag{2.1.45}$$

$$\tilde{E}(\xi,0,t) = -\frac{1}{2}\int_0^{\xi} q(s,t)ds \tag{2.1.46a}$$

$$\tilde{E}(0,\eta,t) = \frac{1}{2}\int_0^{\eta} q(s,t)ds \quad , \tag{2.1.46b}$$

and hence $\tilde{E}(\xi,\eta,t)$ satisfies the Volterra integro-differential equation

$$\begin{aligned}\tilde{E}(\xi,\eta,t) = &-\frac{1}{2}\int_0^{\xi} q(s,t)ds + \frac{1}{2}\int_0^{\eta} q(s,t)ds\\ &+ \int_o^{\eta}\int_o^{\xi} (\tilde{E}_t(\xi,\eta,t) - \tilde{q}(\xi,\eta,t)\tilde{E}(\xi,\eta,t))d\xi d\eta \ .\end{aligned} \tag{2.1.47}$$

The solution of (2.1.47) can formally be obtained by iteration in the form

$$\tilde{E}(\xi,\eta,t) = \sum_{n=1}^{\infty} \tilde{E}_n(\xi,\eta,t) \tag{2.1.48}$$

where

$$\begin{aligned}\tilde{E}_1(\xi,\eta,t) &= -\frac{1}{2}\int_0^{\xi} q(s,t)ds + \frac{1}{2}\int_0^{\eta} q(s,t)ds\\ \tilde{E}_{n+1}(\xi,\eta,t) &= \int_0^{\eta}\int_0^{\xi} (\tilde{E}_{nt}(\xi,\eta,t) - \tilde{q}(\xi,\eta,t)\tilde{E}_n(\xi,\eta,t))d\xi d\eta \ .\end{aligned} \tag{2.1.49}$$

We will now show that the series (2.1.48) converges absolutely and uniformly for (ξ,η,t) on an arbitrary compact subset Ω of $\{(\xi,\eta,t): -x_o < \xi < x_o, -x_o < \eta < x_o, |t| < t_o\}$. To this end let C be a positive constant such that for $(\xi,\eta,t) \in \Omega$ we have with respect to t

$$q(\xi,\eta,t) << C(1-\frac{t}{t_o})^{-1} \tag{2.1.50}$$

Without loss of generality assume $C \geq 1$, $t_o \geq 1$, $x_o \leq 1$. Then from (2.1.49) and the properties of dominants it follows by induction that

$$\tilde{E}_n << \frac{2^n C^n |\xi|^{n-1} |\eta|^{n-1}}{(n-1)!} (1-\frac{t}{t_o})^{-n} \tag{2.1.51}$$

and hence

$$|\tilde{E}_n| \leq \frac{2^n C^n |\xi|^{n-1} |\eta|^{n-1}}{(n-1)!} (1-\frac{t}{t_o})^{-n} \tag{2.1.52}$$

for $(\xi,\eta,t) \in \Omega$. Hence the series (2.1.48) converges absolutely and uniformly for $(\xi,\eta,t) \in \Omega$. In a similar manner it is easily seen that $\tilde{E}(\xi,\eta,t)$ is twice continuously differentiable in Ω, and we can now conclude the existence of the function $\tilde{E}(\xi,\eta,t) = E(s,x,t)$ having the desired properties. Similarly the function $G(s,x,t)$ exists and is twice continuously differentiable in Ω, and we have therefore now established the existence of the operators $\underset{\sim}{T}_1$ and $\underset{\sim}{T}_2$.

We now want to show that if $u(x,t)$ is a classical solution of (2.1.27) in R^+, is continuously differentiable for $0 \leq x < x_o$, $0 < t < t_o$, and satisfies $u(0,t)=0$, then $u(x,t)$ can be represented in the form

$$u(x,t) = \underset{\sim}{T}_1 \{h\} \tag{2.1.53}$$

for some solution $h(x,t)$ of (2.1.39) in R^+ where $h(x,t)$ is continuously differentiable for $0 \leq x < x_o$, $0 < t < t_o$, and satisfies $h(0,t)=0$. (2.1.53) is a Volterra integral equation of the second kind for $h(x,t)$ and hence there exists a solution $h(x,t)$ of (2.1.53) which has the same regularity properties as $u(x,t)$ and satisfies $h(0,t)=u(0,t)=0$. This can be seen by using the resolvent operator to express $h(x,t)$ in terms of $u(x,t)$. To show that this solution of the integral equation (2.1.53) is in fact a solution of the heat equation, we substitute (2.1.53) into (2.1.27) and use (2.1.30), (2.1.31a), (2.1.31b) to obtain

$$\begin{aligned} 0 &= u_{xx} + q(x,t)u - u_t \\ &= (h_{xx}-h_t) + \int_0^x K(s,x,t)(h_{ss}(s,t) - h_t(s,t))ds. \end{aligned} \tag{2.1.54}$$

Since solutions of Volterra integral equations of the second kind are unique, we must have

$$h_{xx} - h_t = 0, \tag{2.1.55}$$

i.e. $h(x,t)$ is a solution of (2.1.39) in R^+.

In a similar manner we can show that if $u(x,t)$ is a classical solution of (2.1.27) in R^+, is continuously differentiable for $0 \leqslant x < x_o$, $0 < t < t_o$, and satisfies $u_x(0,t)=0$, then $u(x,t)$ can be represented in the form $u(x,t)=\underset{\sim}{T}_2\{h\}$ where $h(x,t)$ is a solution of (2.1.39) in R^+ such that $h(x,t)$ is continuously differentiable for $0 \leqslant x < x_o$, $0 < t < t_o$, and satisfies $h_x(0,t)=0$.

We now want to combine the results obtained above to construct an integral operator whose domain and range are independent of the boundary data at $x=0$. Let $u(x,t)$ be a classical solution of (2.1.27) in $R = \{(x,t): -x_o < x < x_o,\ 0 < t < t_o\}$. We will show that there exists a classical solution $h(x,t)$ of (2.1.39) in R such that $u(x,t)$ can be represented in the form

$$u(x,t) = \underset{\sim}{T}_3\{h\} = h(x,t) + \frac{1}{2}\int_{-x}^{x} [K(s,x,t) + M(s,x,t)]h(s,t)ds \tag{2.1.56}$$

(2.1.56) is a Volterra equation of the second kind for $h(x,t)$ and hence can be uniquely solved for $h(x,t)$ where $h(x,t)$ is defined in R. It remains to be shown that $h(x,t)$ is a solution of (2.1.39). From (2.1.39) and (2.1.42) we have that $K(s,x,t) = - K(-s,x,t)$ and $M(s,x,t) = M(-s,x,t)$ and hence we can rewrite (2.1.56) in the form

$$u(x,t) = \frac{1}{2}(h(x,t)-h(-x,t)) + \frac{1}{2}\int_0^x K(s,x,t)\left[h(s,t)-h(-s,t)\right]ds \tag{2.1.57}$$

$$+ \frac{1}{2}(h(x,t) + h(-x,t)) + \frac{1}{2}\int_0^x M(s,x,t)\left[h(s,t)+h(-s,t)\right]ds.$$

Substituting (2.1.57) into (2.1.27) , using (2.1.30), (2.1.31a), (2.1.31b), (2.1.33), (2.1.34a), (2.1.34b) , and rewriting the resulting expression in the form of (2.1.56) gives

$$0 = (h_{xx}-h_t) + \frac{1}{2}\int_{-x}^{x} \left[K(s,x,t)+M(s,x,t)\right](h_{ss}(s,t)-h_t(s,t))ds \tag{2.1.58}$$

Since solutions of Volterra integral equations of the second kind are unique, we can conclude that $h(x,t)$ is a solution of (2.1.39) in R.

We summarize our results in the following theorem :

Theorem 2.1.2 ([10], [12]): Let the coefficient $q(x,t)$ of (2.1.27) be continuously differentiable for $-x_o < x < x_o$, $|t| < t_o$, and an analytic function of t for $|t| < t_o$. Let $R^+ = \{(x,t) : 0 < x < x_o,\ 0 < t < t_o\}$ and $R = \{(x,t) : -x_o < x < x_o,\ 0 < t < t_o\}$.

1) If $u(x,t)$ is a classical solution of (2.1.27) in R^+, continuously differentiable for $0 \leqslant x < x_o, 0 < t < t_o$, and satisfying $u(0,t)=0$, then $u(x,t)$ can be represented in the form $u(x,t) = \underset{\sim}{T}_1\{h\}$ where $h(x,t)$ is a classical solution of (2.1.39) in R^+, continuously differentiable for $0 \leqslant x < x_o$, $0 < t < t_o$, and satisfying $h(0,t)=0$. Conversely for any such $h(x,t)$, $u(x,t) = \underset{\sim}{T}_1\{h\}$ satisfies the above hypothesis on $u(x,t)$.

2) If $u(x,t)$ is a classical solution of (2.1.27) in R^+, continuously differentiable for $0 \leqslant x < x_o$, $0 < t < t_o$, and satisfying $u_x(0,t)=0$, then $u(x,t)$ can be represented in the form $u(x,t) = \underset{\sim}{T}_2\{h\}$ where $h(x,t)$ is a classical solution of (2.1.39) in R^+, continuously differentiable

for $0 \leq x < x_o$, $0 < t < t_o$, and satisfying $h_x(0,t)=0$. Conversely for any such $h(x,t)$, $u(x,t) = \underset{\sim}{T}_2\{h\}$ satisfies the above hypothesis on $u(x,t)$.

3) If $u(x,t)$ is a classical solution of (2.1.27) in R then $u(x,t)$ can be represented in the form $u(x,t) = \underset{\sim}{T}_3\{h\}$ where $h(x,t)$ is a classical solution of (2.1.39) in R. Conversely, for any such $h(x,t)$, $u(x,t)=\underset{\sim}{T}_3\{h\}$ satisfies the above hypothesis on $u(x,t)$.

In the following sections of this chapter we will use the operators $\underset{\sim}{T}_1$ and $\underset{\sim}{T}_2$ to obtain reflection principles for solutions of (2.1.27) and the operator $\underset{\sim}{T}_3$ to construct a complete family of solutions. We also want to show how integral operators can be used to reformulate the first-initial boundary value problem for (2.1.27) as an integral equation in a manner similar to their use in the case of the Dirichlet problem for elliptic equations (c.f. section 1.3). The operators $\underset{\sim}{T}_1$, $\underset{\sim}{T}_2$ and $\underset{\sim}{T}_3$ are not suitable for this purpose since solutions in the range of the operators $\underset{\sim}{T}_1$ and $\underset{\sim}{T}_2$ must satisfy homogeneous boundary data at $x=0$, and solutions in the range of the operator $\underset{\sim}{T}_3$ must be defined in a domain which is symmetric with respect to $x=0$. Hence we will now construct integral operators which are suitable for reformulating the first-initial boundary value problem for (2.1.27) as a Volterra integral equation.

We assume that $q(x,t)$ has been continued in a continuously differentiable manner such that $q(x,t)$ is defined for $-\infty < x < \infty$, $|t| < t_o$, is analytic with respect to t for $|t| < t_o$, and $q(x,t)\equiv 0$ for $|x| \geq a$ where a is a positive constant. We first look for a solution of (2.1.27) for $x \geq -a$, $0 < t < t_o$, in the form

$$u(x,t) = \underset{\sim}{A}_1\{h\} = h(x,t) + \int_x^\infty K^{(1)}(s,x,t)h(s,t)ds \qquad (2.1.59)$$

where $h(x,t)$ is a classical solution of (2.1.29) defined for $x > -a$, $0 < t < t_o$, and $K^{(1)}(s,x,t)$ is a function to be determined which, among other conditions, satisfies

$$K_{-}^{(1)}(s,x,t) \equiv 0 \text{ for } \frac{1}{2}(s+x) \geqslant a. \tag{2.1.60}$$

(2.1.60) guarantees the existence of the integral (2.1.59) for any classical solution $h(x,t)$ of (2.1.29) defined in $x > -a$, $0 < t < t_o$. Substituting (2.1.59) into (2.1.27) and integrating by parts using (2.1.60) shows that (2.1.59) is a solution of (2.1.27) provided $K^{(1)}(s,x,t)$ satisfies (2.1.60) and

$$K_{xx}^{(1)} - K_{ss}^{(1)} + q(x,t)K^{(1)} = K_t^{(1)} \quad ; \qquad s > x \tag{2.1.61}$$

$$K^{(1)}(x,x,t) = -\frac{1}{2}\int_x^\infty q(s,t)ds \quad . \tag{2.1.62}$$

The equations (2.1.60) - (2.1.62) are not enough to uniquely determine $K^{(1)}(s,x,t)$ and so we impose the additional condition

$$K^{(1)}(s,x,t) \equiv 0 \quad \text{for} \quad s < x \; . \tag{2.1.63}$$

We will now construct a function $K^{(1)}(s,x,t)$ satisfying (2.1.60)-(2.1.63) such that $K^{(1)}(s,x,t)$ is twice continuously differentiable with respect to s, x and t for $s \geqslant x$, $|t| < t_o$. In particular this implies that if $h(x,t)$ is a classical solution of (2.1.29) for $x > -a$, $0 < t < t_o$, then $u(x,t)$ as defined by (2.1.59) is a classical solution of (2.1.27) and the domain of regularity of $h(x,t)$ and $u(x,t)$ coincide. Let ξ and η be defined by (2.1.43) and $\tilde{K}^{(1)}(\xi,\eta,t) = K^{(1)}(\xi-\eta, \xi+\eta, t)$. Then (2.1.60)-(2.1.63) become

$$\tilde{K}_{\xi\eta}^{(1)} + q(\xi+\eta,t)\tilde{K}^{(1)} = \tilde{K}_t^{(1)} \quad ; \qquad \eta < 0 \tag{2.1.64}$$

$$\tilde{K}^{(1)}(\xi,0,t) = -\frac{1}{2}\int_\xi^\infty q(s,t)ds \tag{2.1.65}$$

$$\tilde{K}^{(1)}(\xi,\eta,t) \equiv 0 \qquad \text{for} \qquad \xi \geqslant a \tag{2.1.66}$$

$$\tilde{K}^{(1)}(\xi,\eta,t) \equiv 0 \qquad \text{for} \qquad \eta > 0 \quad . \tag{2.1.67}$$

For $\eta \leqslant 0$ we have

$$\tilde{K}^{(1)}(\xi,\eta,t) = -\frac{1}{2}\int_{\xi}^{\infty} q(s,t)ds$$
$$- \int_{\eta}^{\infty}\int_{\xi}^{\infty} \left[q(\alpha+\beta,t)\tilde{K}^{(1)}(\alpha,\beta,t)-\tilde{K}_t^{(1)}(\alpha,\beta,t)\right] d\alpha d\beta. \qquad (2.1.68)$$

Note that (2.1.68) satisfies (2.1.65) since for $\eta=0$ the double integral in (2.1.68) is over the region $\beta \geqslant 0$ where $\tilde{K}^{(1)}(\alpha,\beta,t) \equiv 0$. Now in (2.1.68) make the change of variables

$$\alpha = \frac{\sigma+\mu}{2}$$
$$\beta = \frac{\sigma-\mu}{2} \quad . \qquad (2.1.69)$$

Then (2.1.68) becomes

$$K^{(1)}(s,x,t) = -\frac{1}{2}\int_{\frac{1}{2}(s+x)}^{\infty} q(\sigma,t)d$$
$$-\frac{1}{2}\int_{x}^{\infty}\int_{s+x-\sigma}^{s+\sigma-x} \left[q(\sigma,t)K^{(1)}(\mu,\sigma,t)-K_t^{(1)}(\mu,\sigma,t)\right] d\mu d\sigma \qquad (2.1.70)$$

and from the assumptions on $q(x,t)$ and (2.1.60), (2.1.63) we have that for $\frac{1}{2}(s+x) \leqslant a$, $s \geqslant x$,

$$K^{(1)}(s,x,t) = -\frac{1}{2}\int_{\frac{1}{2}(s+x)}^{a} q(\sigma,t)d\sigma$$
$$-\frac{1}{2}\int_{x}^{\frac{1}{2}(s+x)}\int_{s+x-\sigma}^{s+\sigma-x} \left[q(\sigma,t)K^{(1)}(\mu,\sigma,t)-K_t^{(1)}(\mu,\sigma,t)\right] d\mu d\sigma \qquad (2.1.71)$$
$$-\frac{1}{2}\int_{\frac{1}{2}(s+x)}^{a}\int_{\sigma}^{s+\sigma-x} \left[q(\sigma,t)K^{(1)}(\mu,\sigma,t)-K_t^{(1)}(\mu,\sigma,t)\right] d\mu d\sigma.$$

For $\frac{1}{2}(s+x) < a$, $s \geqslant x$, we now look for a solution of (2.1.71) in the form

$$K^{(1)}(s,x,t) = \sum_{n=1}^{\infty} K_n^{(1)}(s,x,t) \qquad (2.1.72)$$

where

$$K_1^{(1)}(s,x,t) = -\frac{1}{2}\int_{\frac{1}{2}(s+x)}^{a} q(\sigma,t)\,d\sigma \tag{2.1.73}$$

$$K_n^{(1)}(s,x,t) = -\frac{1}{2}\int_x^{\frac{1}{2}(s+x)}\int_{s+x-\sigma}^{s+\sigma-x}\left[q(\sigma,t)K_{n-1}^{(1)}(\mu,\sigma,t) - \frac{\partial}{\partial t}K_{n-1}^{(1)}(\mu,\sigma,t)\right]d\mu\,d\sigma$$

$$-\frac{1}{2}\int_{\frac{1}{2}(s+x)}^{a}\int_{\sigma}^{s+\sigma-x}\left[q(\sigma,t)K_{n-1}^{(1)}(\mu,\sigma,t) - \frac{\partial}{\partial t}K_{n-1}^{(1)}(\mu,\sigma,t)\right]d\mu\,d\sigma$$

for $n \geqslant 2$. Let C be a positive constant such that for $|t| < t_o$

$$q(x,t) << C\left(1 - \frac{t}{t_o}\right)^{-1} \tag{2.1.74}$$

with respect to t, uniformly for $-a \leqslant x \leqslant a$. Without loss of generality we can assume $C \geqslant 1$, $t_o \geqslant 1$. Then

$$K_1^{(1)}(s,x,t) << \frac{1}{2}C\left(1 - \frac{t}{t_o}\right)^{-1}\left(a - \frac{1}{2}(s+x)\right) \tag{2.1.75}$$

and, since in both the double integrals defining $K_2^{(1)}(s,x,t)$ we have $a - \frac{1}{2}(\mu+\sigma) \leqslant a - \frac{1}{2}(s+x)$,

$$K_2^{(1)}(s,x,t) << \frac{C^2}{2}\left(1 - \frac{t}{t_o}\right)^{-2}\left(a-\frac{1}{2}(s+x)\right). \tag{2.1.76}$$

$$\left(\int_x^{\frac{1}{2}(s+x)} 2(\sigma-x)\,d\sigma + \int_{\frac{1}{2}(s+x)}^{a}(s-x)\,d\sigma\right).$$

But in the second integral on the right hand side of (2.1.76) we have $\frac{1}{2}(s+x) \leqslant \sigma$ and hence $s-x \leqslant 2(\sigma-x)$, which implies

$$K_2^{(1)}(s,x,t) << C^2\left(1 - \frac{t}{t_o}\right)^{-2}\left(a-\frac{1}{2}(s+x)\right)\frac{(a-x)^2}{2!}. \tag{2.1.77}$$

Using the identity

$$\int_x^a \frac{(a-\sigma)^{2n}}{(2n)!}(\sigma-x)\,d\sigma = \frac{(a-x)^{2n+2}}{(2n+2)!} \tag{2.1.78}$$

we have by induction that

$$K_n^{(1)}(s,x,t) << C^n(1-\frac{t}{t_o})^{-n}(a-\frac{1}{2}(s+x))(n-1)!\,\frac{(a-x)^{2n}}{(2n)!} \qquad (2.1.79)$$

and hence for $\frac{1}{2}(s+x) \leqslant a, \quad s \geqslant x,$

$$|K_n^{(1)}(s,x,t)| \leqslant C^n(1-\frac{|t|}{t_o})^{-n}(a-\frac{1}{2}(s+x))\,\frac{(n-1)!}{(2n)!}(a-x)^{2n} \qquad (2.1.80)$$

which implies that the series (2.1.72) is absolutely and uniformly convergent for $\frac{1}{2}(s+x) \leqslant a, \quad s \geqslant x \geqslant -a, \; |t| \leqslant t_o - \varepsilon$ where $\varepsilon > 0$ is arbitrarily small, thus establishing the existence of the kernel $K^{(1)}(s,x,t)$. It can easily be verified that $K^{(1)}(s,x,t)$ is twice continuously differentiable with respect to s,x and t for $s \geqslant x, \; |t| < t_o$. We have thus established the existence of the operator $\underset{\sim}{A}_1$ defined by (2.1.59). It is an easy matter to show that every classical solution u(x,t) of (2.1.27) defined for $x > -a, \quad 0 < t < t_o$ (where $q(x,t)\equiv 0$ for $|x| > a$) can be represented in the form $u(x,t) = \underset{\sim}{A}_1\{h\}$ where h(x,t) is a classical solution of (2.1.29) defined in the same domain as u(x,t). For from (2.1.60) we have that the range of integration in the integral in (2.1.59) is in fact only over the finite interval $x \leqslant s \leqslant 3a$ and the invertibility of the operator $\underset{\sim}{A}_1$ follows from the properties of Volterra integral equations of the second kind in the same manner which we previously showed the invertibility of the operators $\underset{\sim}{T}_1$, $\underset{\sim}{T}_2$ and $\underset{\sim}{T}_3$.

In addition to the operator $\underset{\sim}{A}_1$ we will also need the operator $\underset{\sim}{A}_2$ defined by

$$u(x,t) = \underset{\sim}{A}_2\{h\} = h(x,t) + \int_{-\infty}^{x} K^{(2)}(s,x,t)h(s,t)ds \qquad (2.1.81)$$

which maps classical solutions of (2.1.29) defined for $x < a, \quad 0 < t < t_o$, onto classical solutions of (2.1.27) defined in the same domain, where $K^{(2)}(s,x,t)$ is the unique solution of

$$K_{xx}^{(2)} - K_{ss}^{(2)} + q(x,t)K^{(2)} = K_t^{(2)} \quad ; \quad s < x \tag{2.1.82}$$

$$K^{(2)}(x,x,t) = -\frac{1}{2}\int_{-\infty}^{x} q(s,t)ds \tag{2.1.83}$$

$$K^{(2)}(s,x,t) \equiv 0 \text{ for } \tfrac{1}{2}(s+x) \leqslant -a \tag{2.1.84}$$

$$K^{(2)}(s,x,t) \equiv 0 \text{ for } s > x \, , \tag{2.1.85}$$

and $K^{(2)}(s,x,t)$ is twice continuously differentiable with respect to s,x and t for $s \leqslant x$, $|t| < t_o$. The existence of the kernel $K^{(2)}(s,x,t)$ (and hence the operator $\underset{\sim}{A}_2$) follows in the same manner as that of $K^{(1)}(s,x,t)$, where, for $\frac{1}{2}(s+x) \leqslant -a$, $s \leqslant x$, $K^{(2)}(s,x,t)$ satisfies the integro-differential equation

$$\begin{aligned} K^{(2)}(s,x,t) = & -\frac{1}{2}\int_{-a}^{\frac{1}{2}(s+x)} q(\sigma,t)d\sigma \\ & -\frac{1}{2}\int_{\frac{1}{2}(s+x)}^{x}\int_{s+\sigma-x}^{s+x-\sigma}\left[q(\sigma,t)K^{(2)}(\mu,\sigma,t)-K_t^{(2)}(\mu,\sigma,t)\right]d\mu d\sigma \\ & -\frac{1}{2}\int_{-a}^{\frac{1}{2}(s+x)}\int_{s+\sigma-x}^{\sigma}\left[q(\sigma,t)K^{(2)}(\mu,\sigma,t)-K_t^{(2)}(\mu,\sigma,t)\right]d\mu d\sigma \end{aligned} \tag{2.1.86}$$

and

$$K^{(2)}(s,x,t) = \sum_{n=1}^{\infty} K_n^{(2)}(s,x,t) \tag{2.1.87}$$

where

$$\begin{aligned} K_1^{(2)}(s,x,t) = & -\frac{1}{2}\int_{-a}^{\frac{1}{2}(s+x)} q(\sigma,t)d\sigma \\ K_n^{(2)}(s,x,t) = & -\frac{1}{2}\int_{\frac{1}{2}(s+x)}^{x}\int_{s+\sigma-x}^{s+x-\sigma}\left[q(\sigma,t)K_{n-1}^{(2)}(\mu,\sigma,t) - \frac{\partial}{\partial t}K_n^{(2)}(\mu,\sigma,t)\right]d\mu d\sigma \\ & -\frac{1}{2}\int_{-a}^{\frac{1}{2}(s+x)}\int_{s+\sigma-x}^{\sigma}\left[q(\sigma,t)K_{n-1}^{(2)}(\mu,\sigma,t) - \frac{\partial}{\partial t}K_{n-1}^{(2)}(\mu,\sigma,t)\right]d\mu d\sigma \end{aligned} \tag{2.1.88}$$

for $n \geqslant 2$, with

$$|K_n^{(2)}(s,x,t)| \leqslant C^n\left(1-\frac{|t|}{t_o}\right)^{-n}\left(a+\tfrac{1}{2}(s+x)\right)(n-1)!\,\frac{(a+x)^{2n}}{(2n)!} \tag{2.1.89}$$

for $\frac{1}{2}(s+x) \leqslant -a$, $s \leqslant x$. If $u(x,t)$ is a classical solution of (2.1.27) defined for $x < a$, $0 < t < t_o$, then $u(x,t)$ can be represented in the form $u(x,t) = \underset{\sim}{A}_2\{h\}$ where $h(x,t)$ is a classical solution of (2.1.29) defined in the same domain as $u(x,t)$.

We summarize our results in the following theorem:

Theorem 2.1.3 ([15]): Let the coefficient $q(x,t)$ of (2.1.27) be continuously differentiable for $-\infty < x < \infty$, $|t| < t_o$, an analytic function of t for $|t| < t_o$, and such that $q(x,t) \equiv 0$ for $|x| > a$ where a is a positive constant.

1) If $u(x,t)$ is a classical solution of (2.1.27) for $x \geqslant -a$, $0 < t < t_o$, then $u(x,t)$ can be represented in the form $u(x,t) = \underset{\sim}{A}_1\{h\}$ where $h(x,t)$ is a classical solution of (2.1.39) for $x \geqslant -a$, $0 < t < t_o$. Conversely, for any such $h(x,t)$, $u(x,t) = \underset{\sim}{A}_1\{h\}$ satisfies the above hypothesis on $u(x,t)$.

2) If $u(x,t)$ is a classical solution of (2.1.27) for $x \leqslant a$, $0 < t < t_o$, then $u(x,t)$ can be represented in the form $u(x,t) = \underset{\sim}{A}_2\{h\}$ where $h(x,t)$ is a classical solution of (2.1.39) for $x \leqslant a$, $0 < t < t_o$. Conversely, for any such $h(x,t)$, $u(x,t) = \underset{\sim}{A}_2\{h\}$ satisfies the above hypothesis on $u(x,t)$.

2.2 Reflection Principles.

In this section we will use the integral operators constructed in the previous section to obtain reflection principles for solutions of parabolic equations in one space variable. Such results will be needed later on in this chapter to help construct a complete family of solutions for parabolic equations in a manner somewhat similar to the use of the Bergman-Vekua theorem in constructing a complete family of solutions for elliptic equations. Without loss of generality we consider equations written in the canonical form

$$L(u) \equiv u_{xx} + q(x,t)u - u_t = 0 \qquad (2.2.1)$$

defined in a domain $D^+ = \{(x,t) : x_1(t) < x < x_2(t), \quad 0 < t < t_o\}$ where t_o is a positive constant and $x_1(t)$ and $x_2(t)$ are given functions of t. We are primarily interested in the following problem: Let $u(x,t)$ be a solution of $L(u) = 0$ in D^+ such that $u(x,t) \in C^2(D^+) \cap C^1(\bar{D}^+)$ and satisfies the boundary condition

$$\underset{\sim}{I}(u) \equiv a_1(t)u(x_1(t),t) + a_2(t)u_x(x_1(t),t) = g(t) \qquad (2.2.2)$$

on $\sigma : x = x_1(t)$ where $a_1(t), a_2(t)$ and $g(t)$ are given functions of t. Let D^- be the mirror image of D^+ reflected across the arc σ. Under what conditions can $u(x,t)$ be uniquely continued as a solution of (2.2.1) into $D^+ \cup D^- \cup \sigma$? By making the change of variables

$$\xi = x - x_1(t)$$
$$\tau = t \qquad (2.2.3)$$

and following this by a change of variables of the form (2.1.2) we can reduce this problem to the case when $x_1(t)=0$, i.e. D^+ is replaced by the domain $R^+=\{(x,t) : 0 < x < x(t), \quad 0 < t < t_o\}$ where $x(t)$ is a given function of t and (2.2.2) becomes

$$I(u) \equiv \alpha(t)u(0,t) + \beta(t)u_x(0,t) = f(t) \qquad (2.2.4)$$

on $\sigma : x=0$ with $\alpha(t), \beta(t)$ and $f(t)$ given functions of t.

Remark: Even when (2.2.1) is originally the heat equation, the change of variables (2.2.3), (2.1.2) changes this equation into one of the form $L(u)=0$ but where $q(x,t)$ now depends on x and t. Furthermore, $a_1(t) \equiv 0$ does not imply that $\alpha(t) \equiv 0$.

Let R^- be the mirror image of R^+ reflected across $\sigma: x=0$

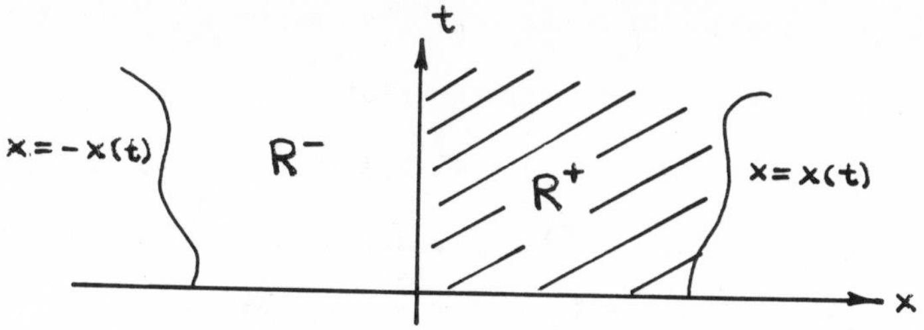

and let $x_o > 0$ be such that $R^+ \cup R^- \cup \sigma \subset B = \{(x,t): -x_o \leq x \leq x_o, |t| < t_o\}$.
We make the following assumptions:

1. $q(x,t) \in C^1(B)$ and is an analytic function of t for $|t| < t_o$ (In particular this implies that $x_1(t)$ should have been analytic for $|t| < t_o$).
2. $\alpha(t)$, $\beta(t)$ and $f(t)$ are analytic for $|t| < t_o$ (This implies that $a_1(t)$, $a_2(t)$ and $g(t)$ should have been analytic for $|t| < t_o$).

We first need to obtain two results on the continuation of solutions to the heat equation

$$h_{xx} = h_t \quad . \tag{2.2.5}$$

The first theorem is the well known reflection principle for solutions of (2.2.5) satisfying homogeneous Dirichlet or Neumann data on x=0, and can be proved in the same manner as the reflection principle for solutions to Laplace equation (c.f. [21]) if one uses the Green (or Neumann) function for the heat equation in a rectangle instead of the Green (or Neumann) function for Laplace's equation in a disc:

Theorem 2.2.1: 1) Let $h(x,t) \in C^2(R^+) \cap C^0(R^+ \cup \sigma)$ be a solution of (2.2.5) in R^+ such that $h(0,t)=0$. Then $h(x,t)$ can be uniquely continued into $R^+ \cup R^- \cup \sigma$ as a solution of (2.2.5) by the rule $h(x,t) = -h(-x,t)$.
2) Let $h(x,t) \in C^2(R^+) \cap C^1(R^+ \cup \sigma)$ be a solution of (2.2.5) in R^+ such that $h_x(0,t)=0$. Then $h(x,t)$ can be uniquely continued into $R^+ \cup R^- \cup \sigma$ as

a solution of (2.2.5) by the rule $h(x,t) = h(-x,t)$.

Our next result is concerned with the analytic continuation of solutions to (2.2.5) satisfying analytic Cauchy data on $x=0$:

Theorem 2.2.2 ([9],[36]): Let $h(x,t) \in C^2(R^+) \cap C^1(R^+ \cup \sigma)$ be a solution of (2.2.5) in R^+ such that $h(0,t)$ and $h_x(0,t)$ are analytic for $|t| < t_0$. Then $h(x,t)$ can be uniquely continued into $-\infty < x < \infty$, $-t_0 < t < t_0$, as a solution of (2.2.5) that is an analytic function of x and t for $|x| < \infty$, $|t| < t_0$.

Proof: From the results of section 2.1 and Holmgren's uniqueness theorem (c.f. [29]) we can represent $h(x,t)$ in the form

$$h(x,t) = -\frac{1}{2\pi i}\oint_{|t-\tau|=\delta} E^{(1)}(x,t,\tau)h(0,\tau)d\tau - \frac{1}{2\pi i}\oint_{|t-\tau|=\delta} E^{(2)}(x,t,\tau)h_x(0,\tau)d\tau \quad (2.2.6)$$

where

$$E^{(1)}(x,t,\tau) = \frac{1}{t-\tau} + \sum_{j=1}^{\infty} \frac{x^{2j}(-1)^j j!}{(2j)!(t-\tau)^{j+1}}$$

$$(2.2.7)$$

$$E^{(2)}(x,t,\tau) = \frac{x}{t-\tau} + \sum_{j=1}^{\infty} \frac{x^{2j+1}(-1)^j j!}{(2j+1)!(t-\tau)^{j+1}}$$

and $\delta > 0$. The statement of the theorem now follows from (2.2.6) and the analyticity of $E^{(1)}(x,t,\tau)$ and $E^{(2)}(x,t,\tau)$.

We can now prove the following reflection principle for solutions of $L(u)=0$ in R^+ satisfying $I(u)=f(t)$ on σ:

Theorem 2.2.3 ([10],[11]): Let $q(x,t)$, $\alpha(t)$, $\beta(t)$ and $f(t)$ satisfy the assumptions stated previously and let $u(x,t) \in C^2(R^+) \cap C^1(R^+ \cup \sigma)$ be a solution of $L(u)= 0$ in R^+ such that $I(u)=f(t)$ on σ: $x=0$. Suppose one of the following three conditions is satisfied:

1) $\beta(t) \equiv 0$, $\alpha(t) \not\equiv 0$ for $|t| < t_o$

2) $\alpha(t) \equiv 0$, $\beta(t) \not\equiv 0$ for $|t| < t_o$

3) $\alpha(t) \not\equiv 0$ and $\beta(t) \not\equiv 0$ for $|t| < t_o$.

Then $u(x,t)$ can be uniquely continued into $R^+ \cup R^- \cup \sigma$ as a solution of $L(u)=0$.

Remark: Note that since $q(x,t)$ is in general not an analytic function of x, the continuation stated above is in general not an analytic continuation with respect to x.

Proof: We first assume condition 1) holds, i.e. $\beta(t)$ is identically zero and $\alpha(t) \not\equiv 0$ in the disc $|t| < t_o$ in the complex t plane. Let $h(x,t)$ be the analytic solution of (2.2.5) given by (2.2.6) where $h(0,t)=f(t)/\alpha(t)$ and $h_x(0,t)=0$, and define the solution $v(x,t)$ of $L(u)=0$ by $v(x,t)=\underset{\sim}{T}_2\{h\}$ where $\underset{\sim}{T}_2$ is the operator constructed in section 2.1 (c.f. Theorem 2.1.2). By construction the operator $\underset{\sim}{T}_2$ preserves Cauchy data on σ: $x=0$ and hence from Theorem 2.2.2 and the regularity of the kernel $K(s,x,t)$ of $\underset{\sim}{T}_2$ we can conclude that $v(x,t)$ is a solution of $L(u)=0$ in $R^+ \cup R^- \cup \sigma$ such that $I(v)=f(t)$ on σ. Hence (by considering $u-v$ instead of u) we can assume without loss of generality that $f(t)\equiv 0$, i.e. $u(0,t)=0$. But now from Theorem 2.1.2 we can represent $u(x,t)$ in the form $u(x,t) = \underset{\sim}{T}_1\{h\}$ where $h(x,t)$ is a solution of (2.2.5) and $h(0,t)=0$, and from Theorem 2.2.1 $h(x,t)$ can be uniquely continued into $R^+ \cup R^- \cup \sigma$ as a solution of (2.2.5). This implies that $u(x,t)$ can be continued as a solution of $L(u)=0$ into $R^+ \cup R^- \cup \sigma$. The uniqueness of the continuation follows from the invertibility of the operator $\underset{\sim}{T}_1$ and the uniqueness of the continuation of $h(x,t)$.

The proof of the theorem under the hypothesis that condition 2) holds proceeds in the same manner by appropriate use of the operators $\underset{\sim}{T}_1$ and $\underset{\sim}{T}_2$.

The proof of the theorem under the hypothesis of condition 3) is a bit more involved. Let $a(t) = \frac{\beta(t)}{\alpha(t)}$ and again assume without loss of generality that $f(t) \equiv 0$. Then the boundary condition (2.2.4) becomes

$$u(0,t) + a(t)u_x(0,t) = 0 \quad . \tag{2.2.8}$$

Let $h^{(1)}(x,t)$ (for $x \geqslant 0$) be the unique solution of the Volterra integral equation

$$\int_0^x u(s,t)ds = -a(t)h^{(1)}(x,t) + \int_0^x \Gamma(s,x,t)h^{(1)}(s,t)ds \tag{2.2.9}$$

where

$$\Gamma(s,x,t) = 1-a(t)K^{(2)}(s,s,t) + \int_s^x K^{(1)}(s,\xi,t)d\xi + a(t)\int_s^x K_s^{(2)}(s,\xi,t)d\xi \tag{2.2.10}$$

and $K^{(1)}(s,x,t), K^{(2)}(s,x,t)$ are the solutions of the initial value problems

$$K_{xx}^{(1)} - K_{ss}^{(1)} + q(x,t)K^{(1)} = K_t^{(1)} \tag{2.2.11}$$

$$K^{(1)}(x,x,t) = -\frac{1}{2}\int_0^x q(s,t)ds \tag{2.2.12a}$$

$$K^{(1)}(0,x,t) = 0 \tag{2.2.12b}$$

and

$$K_{xx}^{(2)} - K_{ss}^{(2)} + \left(q(x,t) - \frac{\dot{a}(t)}{a(t)}\right)K^{(2)} = K_t^{(2)} \tag{2.2.13}$$

$$K^{(2)}(x,x,t) = -\frac{1}{2}\int_0^x \left(q(s,t) - \frac{\dot{a}(t)}{a(t)}\right)ds \tag{2.2.14a}$$

$$K_s^{(2)}(0,x,t) = 0 \tag{2.2.14b}$$

respectively. The existence of the kernels $K^{(1)}(s,x,t)$ and $K^{(2)}(s,x,t)$ and the fact that they are twice constinuously differentiable for $-x_0 < x < x_0$, $|t| < t_0$, follows from the analysis of section 2.1. The existence and uniqueness of $h^{(1)}(x,t)$ is assured from the fact that $a(t) \neq 0$ for $0 < t < t_0$. (2.2.9) also implies that $h^{(1)}(0,t)=0$ and $h^{(1)}(x,t)$ is twice continuously differentiable in $R^+ \cup \sigma$, and three times differentiable in R^+. Differentiating (2.2.9) with respect to x and integrating by parts gives

$$u(x,t) = h^{(1)}(x,t) + h^{(2)}(x,t) + \int_0^x K^{(1)}(s,x,t)h^{(1)}(s,t)ds + \int_0^x K^{(2)}(s,x,t)h^{(2)}(s,t)ds \quad (2.2.15)$$

where $h^{(2)}(x,t)$ is defined by

$$h^{(2)}(x,t) = -a(t)h_x^{(1)}(x,t) \quad . \quad (2.2.16)$$

The fact that $u(x,t)$ satisfies (2.2.8) implies that $h_x^{(2)}(0,t) = -a(t)h_{xx}^{(1)}(0,t) = 0$, i.e. $h_{xx}^{(1)}(0,t) = 0$ for $0 < t < t_o$ since $a(t) \neq 0$ in this interval. Applying the differential equation (2.2.1) to both sides of (2.2.15) and using equations (2.2.11)-(2.2.14) to integrate by parts gives

$$\begin{aligned} 0 = (h_{xx}^{(1)}(x,t)-h_t^{(1)}(x,t)) &-a(t)(h_{xxx}^{(1)}(x,t) -h_{xt}^{(1)}(x,t)) \\ &-a(t)K^{(2)}(x,x,t).(h_{xx}^{(1)}(x,t)-h_t^{(1)}(x,t)) \\ &+ \int_0^x K^{(1)}(s,x,t)(h_{xx}^{(1)}(s,t)-h_t^{(1)}(s,t))ds \\ &+ \int_0^x a(t)K_s^{(2)}(s,x,t)(h_{ss}^{(1)}(s,t)-h_t^{(1)}(s,t))ds. \end{aligned} \quad (2.2.17)$$

Integrating both sides of (2.2.17) with respect to x gives

$$0 = -a(t)(h_{xx}^{(1)}(x,t)-h_t^{(1)}(x,t)) + \int_0^x \Gamma(s,x,t)(h_{ss}^{(1)}(s,t) - h_t^{(1)}(s,t))ds \quad (2.2.8)$$

where $\Gamma(s,x,t)$ is defined by (2.2.10). Since $a(t) \neq 0$ and solutions of nonsingular Volterra integral equations of the second kind are unique, we can conclude that $h^{(1)}(x,t)$ must be a classical solution of (2.2.5) in R^+. From theorem 2.2.1 we can conclude that $h^{(1)}(x,t)$ can be uniquely continued into $R^+ \cup R^- \cup \sigma$ as a solution of (2.2.5) and hence so can $h^{(2)}(x,t)$ as defined by (2.2.16). Equations (2.2.15), (2.2.18) and the regulatiry of $K^{(1)}(s,x,t)$ and $K^{(2)}(s,x,t)$ for $-x_o < s < x_o$, $-x_o < x < x_o$, $|t| < t_o$, now imply that $u(x,t)$ can be continued into $R^+ \cup R^- \cup \sigma$ as a solution of $L(u)=0$. The uniqueness of the continuation follows from the uniqueness of the continuation of $h^{(1)}(x,t)$. The proof of Theorem 2.2.3 is now complete.

2.3 Initial-Boundary Value Problems.

We will now use the integral operators and reflection principles obtained in the last two sections to derive constructive methods for solving initial-boundary value problems for parabolic equations in one space variable defined in domains with time dependent boundaries. Without loss of generality we again consider equations written in the canonical form

$$L(u) \equiv u_{xx} - q(x,t)u - u_t = 0 \qquad (2.3.1)$$

and make the assumption that $q(x,t)$ is continuously differentiable for $-\infty < x < \infty$, $|t| \leqslant t_o$ (where t_o is a positive constant), is analytic with respect to t for $|t| \leqslant t_o$, and $q(x,t)\equiv 0$ for $|x| > a$.

Our first aim is to use the operators $\underset{\sim}{A}_1$ and $\underset{\sim}{A}_2$ of section 2.1 to construct a classical solution $u(x,t)$ of (2.3.1) in the domain $D = \{(x,t): x_1(t) < x < x_2(t),\ 0 < t < t_o\}$ such that $u(x,t)$ is continuous in $\bar{D} = \{(x,t): x_1(t) \leqslant x \leqslant x_2(t),\ 0 \leqslant t \leqslant t_o\}$ and satisfies the initial-boundary data

$$\begin{aligned} u(x_1(t),t) &= \psi_1(t) \quad ; \quad 0 \leqslant t \leqslant t_o \\ u(x_2(t),t) &= \psi_2(t) \quad ; \quad 0 \leqslant t \leqslant t_o \\ u(x,0) &= \psi(x) \quad ; \quad x_1(0) \leqslant x \leqslant x_2(0) \end{aligned} \qquad (2.3.2)$$

where $\psi_1(0)=\phi(x_1(0))$ $\psi_2(0) = \phi(x_2(0))$ ([15]). We will assume that $x_1(t)$ and $x_2(t)$ are continuously differentiable for $0 \leqslant t \leqslant t_o$ and that there exist constants α and β such that $x_1(t) \leqslant \alpha < \beta \leqslant x_2(t)$ for $0 \leqslant t \leqslant t_o$.

We also assume without loss of generality that $\bar{D} \subset \bar{R}$ where $\bar{R} = \{(x,t): -a \leqslant x \leqslant a, \quad 0 \leqslant t \leqslant t_o\}$.

We first reduce the initial-boundary value problem (2.3.1), (2.3.2) to a problem of the same form but with $\phi(x)=0$. To do this it suffices to construct a particular solution $v(x,t)$ of (2.3.1) such that $v(x,t)$ is a classical solution of (2.3.1) for $-\infty < x < \infty$, $0 < t < t_o$, continuous for $-\infty < x < \infty$, $0 \leqslant t \leqslant t_o$, and satisfies

$$v(x,0) = \phi(x); \quad x_1(0) \leqslant x \leqslant x_2(0) \ , \tag{2.3.3}$$

since in this case the reduced problem can be obtained by considering $u(x,t)-v(x,t)$. We look for $v(x,t)$ in the form

$$v(x,t) = \underset{\sim}{A}_1\{h\} = h(x,t) + \int_x^\infty K^{(1)}(s,x,t)h(s,t)ds \tag{2.3.4}$$

where $h(x,t)$ is a solution of

$$h_{xx} = h_t \quad . \tag{2.3.5}$$

To this end we continue $\phi(x)$ in an arbitrary but continuous manner such that $\phi(a)=0$ and define

$$h(x,0) = 0 \ ; \quad x \geqslant a \quad . \tag{2.3.6}$$

Then for $x_1(0) \leqslant x \leqslant a$ let $h(x,0)$ be the unique solution of the Volterra integral equation

$$\phi(x) = h(x,0) + \int_x^a K^{(1)}(s,x,0)h(s,0)ds. \tag{2.3.7}$$

Note that from (2.3.7) we have $h(a,0)=0$ which agrees with (2.3.6). Now for $x \leqslant x_1(0)$ continue $h(x,0)$ in an arbitrary but continuous manner such that

$$h(x,0) = 0 \ ; \quad x \leqslant -a \ . \tag{2.3.8}$$

If we now define $h(x,t)$ by means of the Poisson formula for the heat equation

$$h(x,t) = \frac{1}{2\sqrt{\pi}} \int_{-a}^a \frac{1}{\sqrt{t}} \exp\left[-\frac{(\xi-x)^2}{4t}\right] h(\xi,0)d\xi \ , \tag{2.3.9}$$

it is seen that (2.3.4) satisfies the required conditions for $v(x,t)$. Hence without loss of generality we now consider the initial-boundary value problem (2.3.1),(2.3.2) with $\phi(x)=0$. We will call this problem the reduced initial-boundary value problem for (2.3.1).

We look for a solution of the reduced problem in the form

$$u(x,t) = \underset{\sim}{A}_1\{h^{(1)}\} + \underset{\sim}{A}_2\{h^{(2)}\}$$

$$= h^{(1)}(x,t) + h^{(2)}(x,t) + \int_x^{\infty} K^{(1)}(s,x,t)h^{(1)}(s,t)ds \qquad (2.3.10)$$

$$+ \int_{-\infty}^{x} K^{(2)}(s,x,t)h^{(2)}(s,t)ds$$

where $h^{(1)}(x,0)=h^{(2)}(x,0)=0$, $h^{(1)}(x,t)$ is a solution of (2.3.5) for $x > x_1(t)$, $0 < t < t_o$, and $h^{(2)}(x,t)$ is a solution of (2.3.5) for $x < x_2(t)$, $0 < t < t_o$. Let

$$G_o(x,t,\xi,\tau) = \frac{1}{\sqrt{4\pi(t-\tau)}} \exp\left[-\frac{(x-\xi)^2}{4(t-\tau)}\right], \qquad (2.3.11)$$

and represent $h^{(1)}(x,t)$ and $h^{(2)}(x,t)$ as heat potentials of the second kind (c.f. [39])

$$h^{(1)}(x,t) = \int_0^t \frac{\partial G_o}{\partial \xi}(x,t,x_1(\tau),\tau)\mu_1(\tau)d\tau$$

$$h^{(2)}(x,t) = \int_0^t \frac{\partial G_o}{\partial \xi}(x,t,x_2(\tau),\tau)\mu_2(\tau)d\tau, \qquad (2.3.12)$$

where $\mu_1(\tau)$ and $\mu_2(\tau)$ are continuous densities to be determined. Substituting (2.3.12) into (2.3.10), interchanging orders of integration, and letting x tend to $x_1(t)$ and x tend to $x_2(t)$ respectively, leads to the following system of Volterra integral equations for $\mu_1(t)$ and $\mu_2(t)$:

$$\mu_1(t) + \int_0^t G^{(1)}(x_1(t),t,x_1(\tau),\tau)\mu_1(\tau)d\tau$$

$$+ \int_0^t G^{(2)}(x_1(t),t,x_2(\tau),\tau)\mu_2(\tau)d\tau = \psi_1(t)$$

$$-\mu_2(t) + \int_0^t G^{(1)}(x_2(t),t,x_1(\tau),\tau)\mu_1(\tau)d\tau \tag{2.3.13}$$

$$+ \int_0^t G^{(2)}(x_2(t),t,x_2(\tau),\tau)\mu_2(\tau)d\tau = \psi_2(t)$$

where

$$G^{(1)}(x,t,\xi,\tau) = \frac{\partial G_o}{\partial \xi}(x,t,\xi,\tau)$$

$$+ \int_x^\infty K^{(1)}(s,x,t) \frac{\partial G_o}{\partial \xi}(s,t,\xi,\tau)ds \tag{2.3.14}$$

$$G^{(2)}(x,t,\xi,\tau) = \frac{\partial G_o}{\partial \xi}(x,t,\xi,\tau)$$

$$+ \int_{-\infty}^x K^{(2)}(s,x,t) \frac{\partial G_o}{\partial \xi}(s,t,\xi,\tau)ds.$$

In the derivation of (2.3.13) we have made use of the discontinuity properties of heat potentials of the second kind (c.f. [39]) and the fact that the integrals

$$\int_0^t \int_x^\infty K^{(1)}(s,x,t) \frac{\partial G_o}{\partial \xi}(s,t,\ x_1(\tau),\tau)\mu_1(\tau)dsd\tau \tag{2.3.15}$$

$$\int_0^t \int_{-\infty}^x K^{(2)}(s,x,t) \frac{\partial G_o}{\partial \xi}(s,t,\ x_2(\tau),\tau)\mu_2(\tau)dsd\tau$$

are continuous as x tends to $x_1(t)$ or $x_2(t)$. This last statement follows from estimates of the form

$$| \int^{\infty} K^{(1)}(s,x,t) \frac{\partial G_o}{\partial \xi} (s,t,x_1(\tau),\tau) ds |$$

$$\leqslant \text{constant} \int_x^{\infty} \frac{\partial G_o}{\partial \xi} (s,t,x_1(\tau),\tau) ds \qquad (2.3.16)$$

$$= \text{constant } G_o(x,t,x_1(\tau),\tau)$$

$$\frac{\text{constant}}{\sqrt{t-\tau}} ,$$

which implies for example that the first integral in (2.3.15) is uniformly convergent with respect to x.

The system (2.3.13) is of Volterra type of the second kind with continuous right hand side and hence always has a unique set of continuous solutions $\mu_1(t)$ and $\mu_2(t)$. The solution of the reduced initial-boundary value problem for (2.3.1) is now given by equations (2.3.10) and (2.3.12).

We now turn our attention to the problem of constructing an approximate solution to the initial-boundary value problem (2.3.1), (2.3.2). One method, which is immediate from the above analysis, is to construct an approximate set of solutions to the system of Volterra equations (2.3.13) and then substitute this set of approximate densities $\mu_1(t)$, $\mu_2(t)$ into (2.3.10) and (2.3.12). We will now present an alternate method for obtaining an approximate solution to (2.3.1), (2.3.2) based on the use of a complete family of solutions in a manner analogous to the approach used for elliptic equations in Chapter One. The construction of a complete family of solutions for (2.3.1) is accomplished through the use of the operator $\underset{\sim}{T}_3$ obtained in section 2.1 and the application of the reflection principles obtained in section 2.2. In the rest of this section we will assume that the arcs $x_1(t)$ and $x_2(t)$ are analytic for $0 \leqslant t \leqslant t_o$, although through the use of suitable approximation arguements this assumption can be considerably weakened.

We first consider the set $\{h_n(x,t)\}$ of polynomial solutions to the heat equation (2.3.5) defined by

$$h_n(x,t) = n! \sum_{k=0}^{\left[\frac{n}{2}\right]} \frac{x^{n-2k}\, t^k}{(n-2k)!k!} \tag{2.3.17}$$

$$= (-t)^{n/2}\, H_n\left(\frac{x}{(-4t)^{\frac{1}{2}}}\right)$$

where H_n denotes the Hermite polynomial of degree n(c.f. [45]).

Let x_o be a positive constant, $R = \{(x,t) : x_o < x < x_o,\ 0 < t < t_o\}$ and $\bar{R}$ denote the closure of R.

<u>Theorem 2.3.1</u> ([12]): Let h(x,t) be a classical solution of (2.3.5) in R which is continuous in $\bar{R}$. Then given $\varepsilon > 0$ there exist constants $a_o,\ldots,a_N$ such that

$$\max_{(x,t)\varepsilon\bar{R}} \left|h(x,t) - \sum_{n=0}^{N} a_n h_n(x,t)\right| < \varepsilon .$$

<u>Proof</u>: By the Weierstrass approximation theorem and the maximum principle for the heat equation there exists a solution $w_1(x,t)$ of (2.3.5) in R which assumes polynomial initial and boundary data such that

$$\max_{(x,t)\varepsilon\bar{R}} |h(x,t) - w_1(x,t)| < \frac{\varepsilon}{3} . \tag{2.3.18}$$

Let

$$w_1(-x_o,t) = \sum_{m=0}^{M} b_m t^m$$

$$w_1(x_o,t) = \sum_{m=0}^{M} c_m t^m \tag{2.3.19}$$

and look for a solution of (2.3.5) in the form

$$v(x,t) = \sum_{m=0}^{M} v_m(x) t^m \tag{2.3.20}$$

where $v(-x_o,t) = w_1(-x_o,t)$, $v(x_o,t) = w_1(x_o,t)$. Substituting (2.3.20) into (2.3.5) leads to the following recursion scheme for the $v_m(x)$:

$$\frac{d^2 v_M}{dx^2} = 0$$

$$v_M(-x_o) = b_M \quad , \quad v_M(x_o) = c_M \; , \qquad (2.3.21)$$

$$\frac{d^2 v_{m-1}}{dx^2} = m v_m$$

$$v_{m-1}(-x_o) = b_{m-1}, \quad v_{m-1}(x_o) = c_{m-1} \; ,$$

$m=1,2,\ldots M$. Equation (2.3.21) shows that each $v_m(x)$ is a polynomial in x and is uniquely determined. Now consider $w_2(x,t)=w_1(x,t)-v(x,t)$. By the method of separation of variables it is seen that there exist constants $d_1,\ldots d_L$ such that

$$\max_{(x,t)\epsilon\bar{R}} \left|w_2(x,t)- \sum_{\ell=0}^{L} d_\ell \sin \frac{\ell\pi}{2x_o} (x+x_o)\exp(- \frac{\ell^2\pi^2 t}{4x_o^2})\right| < \frac{\varepsilon}{3} \; .$$

Hence there exists a solution $w_3(x,t)$ of (2.3.5) which is an entire function of the complex variables x and t such that

$$\max_{(x,t)\epsilon\bar{R}} |h(x,t) - w_3(x,t)| < \frac{2\varepsilon}{3} \; . \qquad (2.3.23)$$

From Theorem 2.1.1 (see also Theorem 2.2.2) we can represent $w_3(x,t)$ in the form

$$w_3(x,t) = - \frac{1}{2\pi i} \oint_{|t-\tau|=\delta} E^{(1)}(x,t,\tau)w_3(0,\tau)d\tau - \frac{1}{2\pi i} \oint_{|t-\tau|=\delta} E^{(2)}(x,t,\tau)w_{3x}(0,\tau)d\tau \qquad (2.3.24)$$

where

$$E^{(1)}(x,t,\tau) = \frac{1}{t-\tau} + \sum_{j=1}^{\infty} \frac{x^{2j}(-1)^j j!}{(2j)!(t-\tau)^{j+1}}$$
(2.3.25)
$$E^{(2)}(x,t,\tau) = \frac{x}{t-\tau} + \sum_{j=1}^{\infty} \frac{x^{2j+1}(-1)^j j!}{(2j+1)!(t-\tau)^{j+1}} \quad .$$

By truncating the Taylor series for $w_3(0,\tau)$ and $w_{3x}(0,t)$, (2.3.4), (2.3.25) show that there exists an entire solution $w_4(x,t)$ of (2.3.5) satisfying polynomial Cauchy data on x=0 such that

$$\max_{(x,t)\varepsilon\bar{R}} |w_3(x,t) - w_4(x,t)| < \frac{\varepsilon}{3} \ . \tag{2.3.26}$$

But from (2.3.17) and Holmgren's uniqueness theorem it is seen that there exist positive constants $a_o,\dots,a_N$ such that

$$w_4(x,t) = \sum_{n=0}^{N} a_n h_n(x,t) \quad , \tag{2.3.27}$$

and the proof of the theorem follows from (2.3.27), (2.3.26), (2.3.23) and the triangle inequality.

From Theorem 2.3.1, Theorem 2.1.2, and the continuity of the kernel of the operator $\underset{\sim}{T}_3$, we can now immediately arrive at the following theorem:

<u>Theorem 2.3.2</u> ([12]): Let u(x,t) be a classical solution of (2.2.1) in R which is continuous in $\bar{R}$. Then given $\varepsilon > 0$ there exists constants $a_o,\dots a_N$ such that

$$\max_{(x,t)\varepsilon\bar{R}} |u(x,t) - \sum_{n=0}^{N} a_n u_n(x,t)| < \varepsilon$$

where $u_n(x,t) = \underset{\sim}{T}_3\{h_n\}$.

<u>Remark</u>: Observing that $h_{2n}(x,t)$ is an even function of x and $h_{2n+1}(x,t)$ is an odd function of x, it is seen that we can represent the special solutions $u_n(x,t)$ in terms of the operators $\underset{\sim}{T}_1$ and $\underset{\sim}{T}_2$ by

$$u_{2n}(x,t) = \underset{\sim}{T}_2\{h_{2n}\}$$
$$u_{2n+1}(x,t) = \underset{\sim}{T}_1\{h_{2n+1}\} \quad . \tag{2.3.28}$$

Theorem 2.3.2 shows that the set $\{u_n(x,t)\}$ defined by $u_n(x,t) = \underset{\sim}{T}_3\{h_n\}$ is a complete family of solutions for (2.3.1) in a rectangle. We now want to use the reflection principles of section 2.2 to show that the set $\{u_n(x,t)\}$ is complete in $\bar{D} = \{(x,t) : x_1(t) \leq x \leq x_2(t),\ 0 \leq t \leq t_o\}$ under the assumption that $x_1(t)$ and $x_2(t)$ are analytic.

Theorem 2.3.3 ([13]): Let $u(x,t)$ be a classical solution of (2.2.1) in D which is continuous in $\bar{D}$, where $x_1(t)$ and $x_2(t)$ are analytic for $|t| \leq t_o$. Then given $\varepsilon > 0$ there exist constants $a_o,\ldots,a_N$ such that

$$\max_{(x,t)\in\bar{D}} \left|u(x,t) - \sum_{n=0}^{N} a_n u_n(x,t)\right| < \varepsilon \ .$$

Proof: From Theorem 2.3.2 the theorem will be proved if for a given $\varepsilon > 0$ we can construct a solution $w(x,t)$ of (2.2.1) defined in a rectangle $R = \{(x,t): -x_o < x < x_o,\ 0 < t < t_o\}$ such that $w(x,t)$ is continuous in $\bar{R}$, $\bar{D} \subset \bar{R}$, and

$$\max_{(x,t)\in\bar{D}} |u(x,t) - w(x,t)| < \varepsilon \ . \tag{2.3.29}$$

From the existence of a solution to the first initial-boundary value problem for (2.2.1) (c.f. the first part of this section), the maximum principle for parabolic equations, and the Weierstrass approximation theorem, it is seen that there exists a solution $w(x,t)$ of (2.2.1) in D satisfying analytic boundary data on $x = x_1(t)$, $x = x_2(t)$, and $t = 0$ such that (2.3.29) is valid. From Theorem 2.2.3 (after making the change of variables (2.2.3)) and the regularity theorems for solutions to initial-boundary value problems for parabolic equations (c.f. [26]) we can conclude that $w(x,t)$ can be uniquely continued as a solution of (2.2.1) across the arc $x=x_1(t)$ into the region bounded by the characteristics $t=t_o$, $t=0$, and the analytic curves $x = 2x_1(t) - x_2(t)$, $x=x_2(t)$. Applying Theorem 2.2.3 a second time, but this time continuing $w(x,t)$ across the arc $x=x_2(t)$, shows that $w(x,t)$ can be continued into the region bounded by $t=t_o$, $t=0$, $x=2_1(t) - x_2(t)$, and $x=3x_2(t) - 2x_1(t)$. Due to the fact that $x_1(t) < x_2(t)$ for $0 < t < t_o$, it is seen that by repeating the above procedure we can continue $w(x,t)$ into the entire infinite strip $-\infty < x < \infty$, $0 \leq t \leq t_o$, as a solution of (2.2.1). In particular there exists a rectangle $\bar{R} \supset \bar{D}$ into which $w(x,t)$ can be continued, and we have this established the existence of the desired function $w(x,t)$.

A special case of Theorem 2.3.3 is the following Corollary:

Corollary 2.3.1 ([13]): Let $h(x,t)$ be a classical solution of (2.3.5) in D which is continuous in $\bar{D}$, where $x_1(t)$ and $x_2(t)$ are analytic for $|t| \leq t_o$. Then given $\varepsilon > 0$ there exist constants $a_o,\ldots,a_N$ such that

$$\max_{(x,t)\varepsilon\bar{D}} \left|h(x,t) - \sum_{n=0}^{N} a_n h_n(x,t)\right| < \varepsilon .$$

The complete family $\{u_n(x,t)\}$ can be used to approximate the solution of (2.3.1), (2.3.2) on compact subsets of D in the same manner as for elliptic equations. In particular we orthonormalize the set $\{u_n(x,t)\}$ in the L^2 norm over the base and sides of D to obtain the complete set $\{u_n\}$. Let

$$\begin{aligned} c_n = & \int_0^{t_o} \psi_1(t)\phi_n(x_1(t),t)dt + \int_{x_1(0)}^{x_2(0)} \phi(x)\phi_n(x,0)dx \\ & + \int_0^{t_o} \psi_2(t)\phi_n(x_2(t),t)dt \end{aligned} \qquad (2.3.30)$$

and let D_o be a compact subset of D. From the representation of the solution of (2.3.1), (2.3.2) in terms of the Green's function it is seen that if

$$\int_{\partial D\ t=t_o} \left|u= \sum_{n=0}^{N} c_n\phi_n\right|^2 < \varepsilon \qquad (2.3.31)$$

then

$$\max_{(x,t)\varepsilon D_0} \left|u(x,t) - \sum_{n=0}^{N} c_n\phi_n(x,t)\right| < M\,\varepsilon \qquad (2.3.32)$$

where $M=M(D_0)$ is a constant. Hence an approximate solution to (2.3.1), (2.3.2) on compact subsets of D is given by

$$u^N(x,t) = \sum_{n=0}^{N} c_n\phi_n(x,t). \qquad (2.3.33)$$

Since each $\phi_n(x,t)$ is a solution of (2.3.1), error estimates can be found by finding the maximum of $\left|u(x,t) - \sum_{n=0}^{N} c_n\phi_n(x,t)\right|$ on the base and sides of D and applying the maximum principle. A numerical example using this approach can be found in the Appendix to these lectures.

To conclude this section we show how the operator $\underset{\sim}{P}_1$ of section 2.1 can be used to solve the inverse Stefan problem discussed in the Introduction. The (normalized) inverse Stefan problem is to find a solution of

$$h_{xx} = h_t \;; \quad 0 \leqslant x < s(t), \; t > 0 \qquad (2.3.34)$$

such that on the given analytic arc $x=s(t)$ we have

$$\begin{aligned} h(s(t),t) &= 0, \quad t > 0 \\ h_x(s(t),t) &= -\dot{s}(t), \quad t > 0 \;. \end{aligned} \qquad (2.3.35)$$

In the representation (2.1.23) (with $u(x,t) = h(x,t)$)we place the cycle $|t-\tau| =\delta$ on the two dimensional manifold $x=s(t)$ in the space of two complex variables, and note that since $h(s(t),t)=0$ the integral in (2.1.23) which contains $E^{(1)}(x,t,\tau)$ vanishes. We are thus led to the following representation of the solution to the inverse Stefan problem ([14],[36]):

$$h(x,t) = \frac{1}{2\pi i}\oint_{|t-\tau|=\delta} E^{(2)}(x-s(\tau),t,\tau)\dot{s}(\tau)d\tau \;, \qquad (2.3.36)$$

where

$$E^{(2)}(x,t,\tau) = \frac{x}{t-\tau} + \sum_{j=1}^{\infty} \frac{x^{2j+1}(-1)^j j!}{(2j+1)!(t-\tau)^{j+1}} \quad . \tag{2.3.37}$$

Note that if $s(t)$ is analytic for $0 \leqslant t \leqslant t_o$ then $h(x,t)$ is analytic for $-\infty < x < \infty$, $0 \leqslant t \leqslant t_o$, in particular the temperature $\phi(x)=h(0,t)$ can be obtained by simply evaluating (2.3.36) at $x=0$. Computing thee residue in (2.3.36) leads to the following solution of (2.3.34),(2.3.35):

$$h(x,t) = \sum_{n=1}^{\infty} \frac{1}{(2n)!} \frac{\partial^n}{\partial t^n} \left[x-s(t)\right]^{2n} \quad . \tag{2.3.38}$$

For further discussion of this problem see [36].

III Parabolic equations in two space variables

3.1 Integral Operators and the Riemann Function.

We now want to obtain results for parabolic equations in two space variables which are analogous to the theory previously developed for elliptic equations in two independent variables and parabolic equations in one space variable. In the present case new problems are presented since the domains of the integral operators we are about to construct lie in the space of analytic functions of several complex variables as opposed to analytic functions of one complex variable as in the previous chapters. Nevertheless considerable progress can be made in using analytic function theory to develop constructive methods for solving initial-boundary value problems for parabolic equations in two space variables. In this section we begin the development of this theory by constructing integral operators which map analytic functions of two complex variables onto analytic solutions of the general linear second order parabolic equation written in normal form as

$$u_{xx}+u_{yy}+a(x,y,t)u_x+b(x,y,t)u_y+c(x,y,t)u = d(x,y,t)u_t \tag{3.1.1}$$

where the coefficients of (3.1.1) are entire functions of their independent complex variables which are real valued for x,y and t real. At this point no assumptions are made on the positivity of d(x,y,t). Since the analysis of this section is similar to that of sections 1.1 and 1.2, we will make our presentation somewhat briefer than in previous sections.

The change of variables in $\mathbb{C}^2$

$$\begin{aligned} z &= x+iy \\ z^* &= x-iy \end{aligned} \tag{3.1.2}$$

transforms (3.1.1) into the form

$$L[U] \equiv U_{zz^*} + A(z,z^*,t)U_z + B(z,z^*,t)U_{z^*} + C(z,z^*,t)U - D(z,z^*,t)U_t = 0 \tag{3.1.3}$$

where $A = \frac{1}{4}(a+ib)$, $B = \frac{1}{4}(a-ib)$, $C = \frac{1}{4}c$ and $D = \frac{1}{4}d$. Note that the change of variables (3.1.2) is permissible since we are considering analytic solutions of (3.1.1). Note also, however, that in general classical solutions of (3.1) are not analytic, and hence a problem we will eventually have to face is how to apply the results we are about to obtain on analytic solutions of (3.1.1) to the problem of approximating classical solutions of (3.1.1). We now look for solutions of (3.1.3) in the form

$$U(z,z^*,t) = -\frac{1}{2\pi i}\exp\{-\int_0^{z^*}A(z,\sigma,t)d\sigma\}. \tag{3.1.4}$$

$$\cdot \oint_{|t-\tau|=\delta}\int_{-1}^{1}E(z,z^*,t,\tau,s)f(\frac{z}{2}(1-s^2),\tau)\frac{dsd\tau}{\sqrt{1-s^2}}$$

where $\delta > 0$, $f(z,\tau)$ is an analytic function of two complex variables in a neighbourhood of the point $(0,t)$ and $E(z,z^*,t,\tau,s)$ is an (analytic) function to be determined. The first integral in (3.1.4) is an integration in the complex τ plane in a counterclockwise direction about a circle of radius δ with centre at t, and the second integral is an integration over a curvilinear path in the unit disc in the complex s plane joining the points $s=+1$ and $s=-1$. Substituting (3.1.4) into (3.1.3) and integrating by parts shows that $E(z,z^*,t,\tau,s)$ must satisfy the differential equation

$$(1-s^2)E_{z^*s} - \frac{1}{s}E_{z^*} + 2sz(E_{zz^*} + \tilde{B}E_{z^*} + \tilde{C}E - \tilde{D}E_t) = 0 \tag{3.1.5}$$

where $\tilde{B} = B - \int_0^{z^*}A_z d\sigma$, $\tilde{C} = -(A_z+AB-C)$, $\tilde{D} = D$. We now look for a solution of (3.1.5) in the form

$$E(z,z^*,t,\tau,s) = \frac{1}{t-\tau} + \sum_{n=1}^{\infty}\frac{s^{2n}z^n}{(t-\tau)^{n+1}}\int_0^{z^*}Q^{(2n)}(z,\sigma,t,\tau)d\sigma . \tag{3.1.6}$$

Substituting (3.1.6) into (3.1.5) yields the following recursion formula for the $Q^{(2n)}$:

$$Q^{(2)} = -2(t-\tau)\tilde{C} - 2\tilde{D} \tag{3.1.7}$$

$$(2n+1)Q^{(2n+2)} = -2\Big[(t-\tau)Q_z^{(2n)} + (t-\tau)\tilde{B}\,Q^{(2n)} + (t-\tau)\tilde{C}\int_0^{z^*} Q^{(2n)}d\sigma$$

$$+ (n+1)\tilde{D}\int_0^{z^*} Q^{(2n)}d\sigma - (t-\tau)\tilde{D}\int_0^{z^*} Q_t^{(2n)}d\sigma\Big] ,$$

$$n=1,2,\ldots .$$

It is clear from (3.1.7)that each of the $Q^{(2n)}$, $n=1,2,\ldots$, is uniquely determined. In order to show the existence of $E(z,z^*,t,\tau,s)$ it is now necessary to show the convergence of the series (3.1.6) and it is to this end that we first majorize the functions $Q^{(2n)}$. Let r and t_o be arbitrarily large positive numbers and let B_o be a positive constant such that for $|z| < r$, $|z^*| < r$, $|t| < t_o$, we have

$$\tilde{B}(z,z^*,t) \ll \frac{B_o}{(1-\frac{z}{r})(1-\frac{z^*}{r})(1-\frac{t}{t_o})}$$

$$\tilde{C}(z,z^*,t) \ll \frac{B_o}{(1-\frac{z}{r})(1-\frac{z^*}{r})(1-\frac{t}{t_o})} \tag{3.1.8}$$

$$\tilde{D}(z,z^*,t) \ll \frac{B_o}{(1-\frac{z}{r})(1-\frac{z^*}{r})(1-\frac{t}{t_o})}$$

where "<<" denotes domination. We also have thefact that for $|\tau| \leqslant t_o$, $|t| < t_o$,

$$t-\tau \ll t_o(1-\frac{t}{t_o})^{-1} . \tag{3.1.9}$$

In a straightforward manner which is by now familiar, it can be shown by induction that for any $\varepsilon > 0$ and $|z| < r$, $|z^*| < r$, $|t| < t_o$, $|\tau| \leqslant t_o$ we have (with respect to the variables z,z^*,t)

$$Q^{(2n)} << \frac{M_n 2^n t_o^n (1+\varepsilon)^n}{2n-1} (1-\frac{z}{r})^{-(2n-1)} (1-\frac{z^*}{r})^{-(2n-1)} (1-\frac{t}{t_o})^{-(3n-1)} r^{-n}$$

where (3.1.10)

$$M_1 = \frac{rB_o(1+t_o)}{t_o(1+\varepsilon)} \tag{3.1.11}$$

$$M_{n+1} = M_n(1+\varepsilon)^{-1} \{1+ \frac{B_o r}{(2n-1)^2} (2n-1+r+ \frac{4nr}{t_o})\} .$$

Note that for n sufficiently large we have $M_{n+1} \leqslant M_n$, i.e. there exists a positive constant M which is independent of n such that $M_n \leqslant M$ for all n. Now let $\delta_o \geqslant 1$ and $\alpha > 1$ be positive constants such that

$$\begin{array}{ll} |s| \leqslant \delta_o & |z| < \frac{r}{\alpha} \\ |\tau| \leqslant t_o & |z^*| < \frac{r}{\alpha} \\ |t| < \frac{t_o}{2} & \delta_o \leqslant |t-\tau| \end{array} \tag{3.1.12}$$

where r and t_o arbitrarily large (but fixed) positive numbers and δ_o is arbitrarily small (but again fixed). Then from (3.1.10) it is seen that the series (3.1.6) is majorized by the series

$$\frac{1}{\delta_o} + \sum_{n=1}^{\infty} \frac{r M_n 2^{4n-1} s_o^{2n} t_o^n (1+\varepsilon)^n \alpha^{3n-3}}{\delta_o^{n+1} (2n-1)(\alpha-1)^{4n-2}} . \tag{3.1.13}$$

If α is chosen such that $16\, s_o^2\, t_o(1+\varepsilon)\alpha^3 \delta_o^{-1} (\alpha-1)^{-4} < 1$ then the series (3.1.13) is convergent. Since r, t_o and s_o can be arbitrarily large, δ_o arbitrarily small, and ε is independent of r, t_o, s_o and δ_o, we can now conclude that the series (3.1.6) converges absolutely and uniformly on compact subsets of $\{(z,z^*,t,\tau,s):(z,z^*,t,\tau,s)\varepsilon\ \mathbb{C}^5, t\neq\tau\}$, i.e. $E(z,z^*,t,\tau,s)$ exists and is an entire function of its independent complex variables except for an (essential) singularity at $t=\tau$. Note that if the coefficients $a(x,y,t)$, $b(x,y,t)$, $c(x,y,t)$ and $d(x,y,t)$ are independent of t, then $E(z,z^*,t,\tau,s)$ is a function only of z, z^*, s and $t-\tau$, i.e. $E(z,z^*,t,\tau,s) = E(z,z^*,t-\tau,s)$. In particular for the special case of the heat equation

$$u_{xx} + u_{yy} = u_t \tag{3.1.14}$$

we have

$$E(z,z^*,t,\tau,s) = \frac{\pi}{t-\tau} \sum_{n=0}^{\infty} \frac{(-1)^n}{\Gamma(n+\frac{1}{2})} \left(\frac{r^2 s^2}{t-\tau} \right)^n \tag{3.1.15}$$

where $r^2 = zz^* = x^2+y^2$.

We have now shown that the operator $\underset{\sim}{P}_2$ defined by

$$U(z,z^*,t) = \underset{\sim}{P}_2\{f\} = -\frac{1}{2\pi i} \exp\left\{ -\int_0^{z^*} A(z,\sigma,t)d\sigma \right\}.$$

$$\cdot \oint_{|t-\tau|=\delta} \int_{-1}^{1} E(z,z^*,t,\tau,s) f\left(\frac{z}{2}(1-s^2),\tau\right) \frac{ds d\tau}{\sqrt{1-s^2}} \tag{3.1.16}$$

exists and maps analytic functions of two complex variables into the class of (complex valued) solutions of (3.1.3). An elementary power series analysis shows that solutions of (3.1.3) which are real valued for t real and $z^*=\bar{z}$ (i.e. x and y real) are uniquely determined by their values on the characteristic plane $z^*=0$. Furthermore, since the coefficients of (3.1.1) are real valued for x,y and t real, the operator $\mathrm{Re}\underset{\sim}{P}_2\{f\}$ (where "Re" denotes "take the real part") defines a real valued solution of (3.1.1) provided we set $z^*=\bar{z}$ and keep t real. Evaulating $\mathrm{Re}\, \underset{\sim}{P}_2\{f\}$ at $z^*=0$ and keeping t real gives

$$U(z,0,t) = \mathrm{Re}\, \underset{\sim}{P}_2\{f\}\Big|_{z^*=0} \tag{3.1.17}$$

$$= -\frac{1}{4\pi i} \oint_{|t-\tau|=\delta} \int_{-1}^{1} \left[f\left(\frac{z}{2}(1-s^2),\tau\right) + \bar{f}(0,\tau)\exp\left(-\int_0^z \bar{A}(0,\sigma,t)d\sigma\right)\right] \cdot$$

$$\cdot \frac{ds d\tau}{(t-\tau)\sqrt{1-s^2}}$$

$$= \frac{1}{2}\int_{-1}^{1} f(\tfrac{z}{2}(1-s^2),t)\,\frac{ds}{\sqrt{1-s^2}} + \frac{\pi}{2}\,\bar{f}(0,t)\exp(-\int_{o}^{z}\bar{A}(0,\sigma,t)d\sigma)$$

where $\bar{f}(z,t) = \overline{f(\bar{z},\bar{t})}$ and $\bar{A}(z,z^*t) = \overline{A(\bar{z},\bar{z}^*,\bar{t})}$. A solution of the integral equation (3.1.17) is given by (c.f. section 1.2)

$$f(\tfrac{z}{2},t) = -\frac{1}{2\pi i}\int_{\gamma}\Big[2U(z(1-s^2),0,t) - U(0,0,t)\exp(-\int_{o}^{z}\bar{A}(0,\sigma,t)d\sigma)\Big]\frac{ds}{s^2} \tag{3.1.18}$$

where γ is a rectifiable arc joining the points $s=-1$ and $s=+1$ and not passing through the origin. (3.1.17) and (3.1.18) show that if $U(z,\bar{z},t) = u(x,y,t)$ is real valued for x,y and t real, then $f(z,t)$ can be chosen such that $U(z,0,t)$ assumes prescribed (analytic) values. We summarize our results in the following theorem:

<u>Theorem 3.1.1</u> ([16], [17]): Let $u(x,y,t)$ be a real valued analytic solution of (3.1.1) defined in some neighbourhood of the point $(0,0,t_o)$. Then $u(x,y,t) = U(z,\bar{z},t)$ can be represented in the form $u(x,y,t) = \mathrm{Re}\,\underset{\sim}{P}_2\{f\}$ where $f(z,t)$ is an analytic function of z and t in some neighbourhood of the point $(0,t_o)$. Conversely, for every analytic function $f(z,t)$ defined in some neighbourhood of the point $(0,t_o)$, $u(x,y,t) = \mathrm{Re}\,\underset{\sim}{P}_2\{f\}$ defines a real valued analytic solution of (3.1.1) in some neighbourhood of the point $(0,0,t_o)$.

The operator $\underset{\sim}{P}_2$ is in fact closely related to the Bergman operator $\underset{\sim}{B}_2$ for elliptic equations in two independent variables constructed in section 1.2. To see this we consider the case in which the coefficients and solution of (3.1.1) are independent of t and hence $u(x,y,t) = u(x,y)$ satisfies the elliptic equation

$$u_{xx}+u_{yy}+a(x,y)u_x+b(x,y)u_y+c(x,y)u = 0\,. \tag{3.1.19}$$

In this situation the associated analytic function $f(z,t) = f(z)$ is independent of t, and taking the real part of (3.1.16) and integrating termwise yields the representation

$$U(z,\bar{z}) = \mathrm{Re} \Big[\exp \{- \int_0^{\bar{z}} A(z,\sigma)d\sigma\}. \qquad (3.1.20)$$

$$\cdot \int_{-1}^{1} E(z,\bar{z},s) f(\tfrac{z}{2}(1-s^2)) \frac{ds}{\sqrt{1-s^2}} \Big]$$

where

$$E(z,z^*,s) = 1 + \sum_{n=1}^{\infty} s^{2n} z^n \int_0^{z^*} P^{(2n)}(z,\zeta^*)d\zeta^* \qquad (3.1.21)$$

with the $P^{(2n)}$ being defined recursively by

$$P^{(2)} = -2\tilde{C} \qquad (3.1.22)$$

$$(2n+1)P^{(2n+2)} = -2\Big[P_z^{(2n)} + \tilde{B}P^{(2n)} + \tilde{C} \int_0^{z^*} P^{(2n)} d\zeta^* \Big] ,$$

A comparison of (3.1.20)-(3.1.22) with (1.3.21), (1.2.13) and (1.3.14) shows that the operator defined by (3.1.20) is identical with the Bergman operator Re $\underset{\sim}{B}_2\{f\}$.

In addition to the operator $\underset{\sim}{P}_2$ we will also need to make use of a generalized form of this operator which we will denote by $\underset{\sim}{P}_2^*$ and is defined by

$$U(z,z^*,t) = \underset{\sim}{P}_2^*\{f\} = -\frac{1}{2\pi i} \exp\{ - \int_{\zeta^*}^{z^*} A(z,\sigma,t)d\sigma\}. \qquad (3.1.23)$$

$$\cdot \oint_{|t-t_1|=\delta} \int_{-1}^{1} E^*(z,z^*,t,t_1,s) f(\frac{(z-\zeta)}{2}(1-s^2),t_1) \frac{ds dt_1}{\sqrt{1-s^2}}$$

where $\delta > 0$, $(\zeta,\zeta^*) \in \mathbb{C}^2$, $f(z,t_1)$ is an analytic function of two complex variables in some neighbourhood of the point $(0,t)$, and

$$E^*(z,z^*,t,t_1,s) = \frac{1}{t-t_1} + \sum_{n=1}^{\infty} \frac{s^{2n}(z-\zeta)^n}{(t-t_1)^{n+1}} \int_{\zeta^*}^{z^*} Q^{(2n)}(z,\sigma,t,t_1)d\sigma \qquad (3.1.24)$$

with

$$Q^{(2)} = -2(t-t_1)\tilde{C} - 2\tilde{D}$$

$$(2n+1)Q^{(2n+2)} = -2\Big[(t-t_1)Q_z^{(2n)} + (t-t_1)\tilde{B}\, Q^{(2n)} + (t-t_1)\tilde{C}\int_{\zeta^*}^{z^*} Q^{(2n)} d\sigma \tag{3.1.25}$$

$$+ (n+1)\tilde{D}\int_{\zeta^*}^{z^*} Q^{(2n)} d\sigma - (t-t_1)\tilde{D}\int_{\zeta^*}^{z^*} Q_t^{(2n)} d\sigma\Big] .$$

By slightly modifying our previous analysis for the case of the operator $\underset{\sim}{P}_2$ (c.f. section 1.1) it can be seen that the operator $\underset{\sim}{P}_2^*$ exists and maps analytic functions of two complex variables defined in some neighbourhood of the point $(0,t_o)$ into the class of analytic solutions of (3.1.1) defined in some neighbourhood of the point (ζ,ζ^*,t_o). It is also easy to see that $E^*(z,z^*,t,t_1,s) = E^*(z,z^*,t;\ \zeta,\zeta^*,t_1,s)$ is an entire function of its seven independent complex variables except for an essential singularity at $t=t_1$.

We make the observation that if, as a function of t, $f(z,t)$ has an isolated singularity at $t=\tau$ for a given $\tau \in \mathbb{C}^1$ then $U(z,z^*,t) = \underset{\sim}{P}_2^*\{f\}$ also has an isolated singularity at $t=\tau$.

We will now use the integral operator $\underset{\sim}{P}_2^*$ associated with the adjoint equation to (3.1.13) to construct the Riemann function for (3.1.1). The Riemann function $R(z,z^*,t;\zeta,\zeta^*,\tau)$ for (3.1.1) is defined to be the (unique) solution of the adjoint equation

$$M[V] = V_{zz^*} - \frac{\partial(AV)}{\partial z} - \frac{\partial(BV)}{\partial z^*} + CV + \frac{\partial}{\partial t}(DV) = 0 \tag{3.1.26}$$

satisfying the initial data

$$R(z,\zeta^*,t;\zeta,\zeta^*,\tau) = \frac{1}{t-\tau}\exp\left\{\int_{\zeta}^{z} B(\sigma,\zeta^*,t)d\sigma\right\}$$

$$R(\zeta,z^*,t;\zeta,\zeta^*,\tau) = \frac{1}{t-\tau}\exp\left\{\int_{\zeta^*}^{z^*} A(\zeta,\sigma,t)d\sigma\right\} \tag{3.1.27}$$

(c.f. [17], [37]).

To establish the existence of the Riemann function we let

$f(\frac{z}{2},t) = \frac{1}{t-\tau}F(\frac{z}{2},t)$ where

$$F(\frac{z}{2},t) = -\frac{1}{2\pi}\int_{\gamma}\exp\{\int_{0}^{z(1-\rho^2)} B(\sigma+\zeta,\zeta^*,t)d\sigma\}\frac{d\rho}{\rho^2} \qquad (3.1.28)$$

(with γ defined as in (3.1.18)) and define the solution $V(z,z^*,t;\zeta,\zeta^*,\tau)$ of (3.1.26) by

$$V(z,z^*,t;\zeta,\zeta^*,\tau) = \underset{\sim}{P}_2^*\{\frac{F(z,t)}{t-\tau}\} \qquad (3.1.29)$$

where $\underset{\sim}{P}_2^*$ is the integral operator associated with (3.1.26).

Then from the reciprocal relations (c.f. section 1.3)

$$\int_{-1}^{1} f(\frac{z}{2}(1-s^2))\frac{ds}{\sqrt{1-s^2}} = g(z) \qquad (3.1.30)$$

$$-\frac{1}{2\pi}\int_{\gamma} g(z(1-\rho^2)\frac{d\rho}{\rho^2} = f(\frac{z}{2})$$

we have that

$$V(z,\zeta^*,t;\zeta,\zeta^*,\tau) = \frac{1}{t-\tau}\exp\{\int_{0}^{(z-\zeta)} B(\sigma+\zeta,\zeta^*,t)d\sigma\}$$

$$= \frac{1}{t-\tau}\exp\{\int_{\zeta}^{z} B(\sigma,\zeta^*,t)d\sigma\} \qquad (3.1.31)$$

$$V(\zeta, z^*,t;\zeta,\zeta^*,\tau) = \frac{1}{t-\tau}\exp\{\int_{\zeta^*}^{z^*} A(z,\sigma,t)d\sigma\},$$

i.e. $V(z,z^*,t;\zeta,\zeta^*,\tau)$ is in fact the Riemann function $R(z,z^*,t;\zeta,\zeta^*,\tau)$.

Note that except for an essential singularity at $t=\tau$ the Riemann function is an entire function of its six independent complex variables.

3.2 Complete Families of Solutions.

We will now use the integral operators and Riemann function constructed in the last section to construct a complete family of solutions to (3.1.1) in the space of real valued classical solutions to (3.1.1) defined in a cylinder $Dx(0,T)$ where T is a positive constant. We will assume that D is a bounded, simply connected domain in $\mathbb{R}^2$ containing the origin and that the cofficient $d(x,y,t)$ is greater than zero in $Dx(0,T)$.

Let $u(x,y,t)$ be a real valued classical solution of (3.1.1) in $Dx(0,T)$, $\overline{D_0}$ and $\overline{D_1}$ compact subsets of D such that $D \supset \overline{D_1} \supset \overline{D_0}$ and let ∂D_1 be analytic. From the existence theorems for solutions of initial-boundary value problems for parabolic equations (c.f. [26], [27]) and the maximum principle for parabolic equations we can conclude that for $\varepsilon > 0$, $\delta_0 > 0$, there exists a solution $u_1(x,y,t)$ of (3.1.1) in $D_1x(\frac{\delta_0}{2}, T-\frac{\delta_0}{2})$ such that $u_1(x,y,t)$ is continuous in $\overline{D_1}x[\frac{\delta_0}{2}, T-\frac{\delta_0}{2}]$, assumes analytic Dirichlet data on $\partial D_1x[\frac{\delta_0}{2}, T-\frac{\delta_0}{2}]$, and satisfies

$$\max_{\overline{D_1}x[\frac{\delta_0}{2}, T-\frac{\delta_0}{2}]} |u_1-u| < \varepsilon/2. \tag{3.2.1}$$

From a result of Friedman ([27] p.212) we can conclude that $u_1(x,y,t)$ is analytic in $\overline{D_1}x(\frac{\delta_0}{2}, T-\frac{\delta_0}{2})$, i.e. for every point $(x_0,y_0,t_0) \in \overline{D_1}x(\frac{\delta_0}{2},T-\frac{\delta_0}{2})$ there exists a ball in $\mathbb{C}^3$ with centre at (x_0,y_0,t_0) such that as a function of the complex variables x,y,t, $u_1(x,y,t)$ is analytic in this ball. By standard compactness arguments we can conclude that $u_1(x,y,t)$ is analytic in some "thin" neighbourhood in $\mathbb{C}^3$ of the product domain $\overline{D_1}x[\delta_0,T-\delta_0]$. We now want to show that $U_1(z,z^*,t) = u_1(x,y,t)$ can be analytically continued as a function z,z^* and t (where $z^*=\bar{z}$ for x and y real) into the product domain $D_1xD_1^*xE$ where

$$D_1 = \{z : z \in D_1\} \tag{3.2.2}$$

$$D_1^* = \{z^* : \bar{z}^* \in D_1\}$$

and E is an ellipse in $\mathbb{C}^1$ containing the interval $[\delta_o, T-\delta_o]$ such that for $(x,y)\in\bar{D}_1$, $u(x,y,t)$ is an analytic function of t in E. This result is the analogue for parabolic equations of the Bergman-Vekua theorem for elliptic equations (c.f. section 1.1).

Theorem 3.2.1 ([17]): $U_1(z,z^*,t)$ is analytic in $D_1 x D_1^* x E$.

Proof: From Stokes theorem we have that for u and v analytic in a neighbourhood of $\bar{D}_1 x [\delta_o, T-\delta_o]$

$$\iiint_{D_1 x\Omega} (v\mathcal{L}[u] - u\mathcal{M}[v])\,dxdydt = \iint_{\partial D_1 x\Omega} H[u,v] \tag{3.2.3}$$

where $\mathcal{L}$ is the differential operator defined by (3.1.1), $\mathcal{M}$ is its adjoint, $\Omega = \{t: |t-\tau|=\delta\}$ such that $\Omega \subset E$, and

$$H[u,v] = \{(vu_x - uv_x + auv)dydt - (vu_y - uv_y + buv)dxdt - (duv)dxdy\}\ . \tag{3.2.4}$$

The region of integration $D_1 x\Omega$ in (3.2.3) can be geometrically visualised as a three dimensional torus lying in the six dimensional space $\mathbb{C}^3$. Note that on ∂D_1 we have dxdy=0. Now let D_ε be a small disc of radius ε about the point (ξ,η), $u=u_1(x,y,t)$, $v=R(z,\bar{z},t;\zeta,\bar{\zeta},\tau)\log r$ (where $r^2=(z-\zeta)(\bar{z}-\bar{\zeta})$, $\zeta=\xi+i\eta$, $\bar{\zeta}=\xi-i\eta$) and apply (3.2.3) to u and v with the torus $D_1 x\Omega$ replaced by the hollow torus $D_1 \setminus D_\varepsilon x\Omega$. Letting ε tend to zero now gives

$$0 = \lim_{\varepsilon\to 0} \{ \iint_{\partial(D_1\setminus D_\varepsilon)x\Omega} H[u_1, R\log r] + \iiint_{D_1\setminus D_\varepsilon x\Omega} u_1\mathcal{M}[R\log r]\,dxdydt\}$$

$$= \iint_{\partial D_1 \times \Omega} H[u_1, R\log r] + 2\pi \oint_{\Omega} \frac{u_1(\xi,\eta,t)}{t-\tau}\, dt$$

$$+ \iiint_{D_1 \times \Omega} u_1 \mathcal{M}[R\log r]\, dxdydt \tag{3.2.5}$$

$$= \iint_{\partial D_1 \times \Omega} H[u_1, R\log r] + 4\pi^2 i\, u_1(\xi,\eta,\tau)$$

$$+ \iiint_{D_1 \times \Omega} u_1 \mathcal{M}[R\log r]\, dxdydt$$

i.e.

$$u_1(\xi,\eta,\tau) = \frac{i}{\pi^2}\left(\iint_{\partial D_1 \times \Omega} H[u_1, R\log r] + \iiint_{D_1 \times \Omega} u_1 \mathcal{M}[R\log r]\, dxdydt \right). \tag{3.2.6}$$

Returning now to the complex coordinates z, z^*, we see from the fact that $M[R] = 0$ that

$$\mathcal{M}[R\log r] = M[R\log r] = 2\, \frac{\partial R/\partial z - BR}{\zeta^* - z^*} + 2\, \frac{\partial R/\partial z^* - AR}{\zeta - z}, \tag{3.2.7}$$

and hence from (3.1.27) we have that $\mathcal{M}[R\log r]$ is an entire function of its independent complex variables except for an essential singularity at $t=\tau$. Hence, replacing $\bar{\zeta}$ by ζ^*, we see that the second integral in (3.2.6) can be continued to an entire function of ζ and ζ^* for $\tau \in E$. The first integral in (3.2.6) can be continued to an analytic function of ζ, ζ^* and τ for $(\zeta,\zeta^*,\tau) \in D_1 x D_1^* x E$. Hence (3.2.6) shows that $U_1(\zeta,\zeta^*,\tau) = u_1(\xi,\eta,\tau)$ is analytic in $D_1 x D_1^* x E$ and the theorem is established.

With the help of the above theorem on the analytic continuation of analytic solutions to (3.1.1) we can now establish the following version of Runge's Theorem for parabolic equations in two space variables:

Theorem 3.2.2 ([17]): Let $u(x,y,t)$ be a real valued classical solution of (3.1.1) in $Dx(0,T)$ where $d(x,y,t) > 0$ in $Dx(0,T)$ and let $\overline{D_o} x [\delta_o, T-\delta_o]$ be a compact subset of $Dx(0,T)$. Then for every $\varepsilon > 0$ there exists an entire solution $u_o(x,y,t)$ of (1.1) such that

$$\max_{\bar{D}_o \times [\delta_o, T-\delta_o]} |u-u_o| < \varepsilon \tag{3.2.8}$$

Proof: Let $u_1(x,y,t)$ be an analytic solution of (3.1.1) in $\bar{D}_1 \times [\frac{\delta_o}{2}, T-\frac{\delta_o}{2}]$ such that (3.2.1) is valid. From Theorem 3.2.1 we have that $U_1(z,z^*,t) = u_1(x,y,t)$ is analytic in $D_1 \times D_1^* \times E$, and from Theorem 3.1.1 we can represent $U_1(z,\bar{z},t)$ in this domain in the form $U_1(z,\bar{z},t) = \text{Re}\, \underset{\sim}{P}_2\{f\}$ where $f(z,t)$ is given by (3.1.18) with U replaced by U_1. (We emphasize again the importance of Theorem 3.2.1 which tells us that $U(z,0,t)$ is analytic in $D_1 \times E$. From (3.1.18) this implies $f(\frac{z}{2},t)$ is analytic in $D_1 \times E$ and hence $\text{Re}\, \underset{\sim}{P}_2\{f\}$ is analytic in $D_1 \times D_1^* \times E$ and therefore must equal $U_1(z,\bar{z},t)$ not only locally but in the entire product domain $D_1 \times D_1^* \times E$.) Since product domains are Runge domains of the first kind (c.f. [28], p.49). we can approximate $U_1(z,0,t)$ (and hence $f(\frac{z}{2},t)$) on compact subsets of $D_1 \times E$ by a polynomial. In particular since $\text{Re}\, \underset{\sim}{P}_2\{f\}$ tends to zero as $f(z,t)$ tends to zero in the maximum norm, we can conclude that there exists a polynomial $f_n(z,t)$ and entire solution $u_2(x,y,t) = \text{Re}\, \underset{\sim}{P}_2\{f_n\}$ of (3.1.1) such that

$$\max_{\bar{D}_o \times [\delta_o, T-\delta_o]} |u_2-u_1| < \varepsilon/2 \quad . \tag{3.2.9}$$

The theorem now follows from (3.2.1) and (3.2.9) by the use of the triangle inequality and the fact that $\bar{D}_1 \supset \bar{D}_o$.

As an immediate consequence of Theorem 3.2.2 we have the following corollary, where "Im" denotes "take the imaginary part":

Corollary 3.2.1: Let $u(x,y,t)$ be a real valued classical solution of (3.1.1) in $D \times (0,T)$ where $d(x,y,t) > 0$ in $D \times (0,T)$, $\bar{D}_o \subset D$, $\delta_o > 0$, and let

$$\begin{aligned} u_{2n,m} &= \text{Re}\, \underset{\sim}{P}_2\{z^n t^m\} \\ u_{2n+1,m} &= \text{Im}\, \underset{\sim}{P}_2\{z^n t^m\} \end{aligned} \tag{3.2.10}$$

for $n,m = 0,1,2,\ldots$. Then for any $\varepsilon > 0$ there exists integers $N = N(\varepsilon)$, $M = M(\varepsilon)$, and constants a_{nm}, $n = 0, 1,..,N$, $m = 0,1,\ldots,M$, such that

$$\max_{\bar{D}_0 x[\delta_0, T-\delta_0]} \left| u - \sum_{n=0}^{N} \sum_{m=0}^{M} a_{nm} u_{nm} \right| < \varepsilon . \tag{3.2.11}$$

We note that for the case of the heat equation (3.1.4) we have from (3.1.15) and the result

$$\int_{-1}^{1} (1-s^2)^{n-\frac{1}{2}} s^{2k} ds = \frac{\Gamma(n+\frac{1}{2})\Gamma(k+\frac{1}{2})}{\Gamma(n+k+1)} \tag{3.2.12}$$

that

$$u_{2n,m}(x,y,t) = \cos n\theta \sum_{k=0}^{m} \frac{\pi\ \Gamma(m+1)\Gamma(n+\frac{1}{2})}{\Gamma(k+1)\Gamma(m-k+1)\Gamma(n+k+1)} r^{2k+n} t^{m-k} \qquad (3.2.12)$$

$$u_{2n+1,m}(x,y,t) = \sin n\theta \sum_{k=0}^{m} \frac{\pi\ \Gamma(m+1)\Gamma(n+\frac{1}{2})}{\Gamma(k+1)\Gamma(m-k+1)\Gamma(n+k+1)} r^{2k+n} t^{m-k}$$

where $x=r\cos\theta$, $y=r\sin\theta$. Noting that in this special case $u_{n,m}(x,y,t)$ is a polynomial in x,y and t, it follows from the uniqueness theorem for Cauchy's problem for the heat equation (c.f.[35]) that another complete family of solutions (on compact subsets) for the heat equation defined in $Dx(0,T)$ is given by

$$v_{n,m}(x,y,t) = h_n(x,t)h_m(y,t) \tag{3.2.14}$$

for $n,m = 0,1,2,\ldots$ where $h_n(x,t)$ is the polynomial defined in (2.3.17).

We now consider the case when (3.1.1) is of the form

$$u_{xx} + u_{yy} + c(x,y)u = d(x,y)u_t \tag{3.2.15}$$

and show that if ∂D is three times continuously differentiable, $u(x,y,t)$ is continuous in $\bar{D}x[0,T]$, and $d(x,y) > 0$ in $\bar{D}$, then the family (3.2.10) is in fact complete "up to the boundary", i.e. in (3.2.11) $\bar{D}_0 x[\delta_0, T-\delta_0]$ can be replaced by $\bar{D}x[0,T]$. Equations of the form (3.2.15) are of particular interest since a wide variety of equations appearing in mathematical physics can be written in the form (3.2.15). For example the equation of heat conduction in a nonhomogeneous medium is governed by the equation

$$\frac{\partial}{\partial x}\left(k(x,y)\frac{\partial u}{\partial x}\right) + \frac{\partial}{\partial y}\left(k(x,y)\frac{\partial u}{\partial y}\right) = c(x,y)\frac{\partial u}{\partial t} \qquad (3.2.16)$$

where $k(x,y)$ and $c(x,y)$ are positive, continuous known functions and $u=u(x,y,t)$ denotes the temperature in the medium. Writing (3.2.16) in the form

$$k(x,y)(u_{xx}+u_{yy}) + k_x(x,y)u_x + k_y(x,y)u_y = c(x,y)u_t \qquad (3.2.17)$$

and dividing by $\sqrt{k(x,y)}$ we obtain (after rearrangement)

$$v_{xx}+v_{yy} - \frac{\Delta(\sqrt{k(x,y)})}{\sqrt{k(x,y)}}\, v = \frac{c(x,y)}{\sqrt{k(x,y)}}\, v_t \qquad (3.2.18)$$

where $v(x,y,t) = \sqrt{k(x,y)}\, u(x,y,t)$ and $\Delta = \frac{\partial^2}{\partial x^2} + \frac{\partial^2}{\partial y^2}$.

Now let $u(x,y,t)$ be a classical real valued solution of (3.2.15) in a cylindrical domain $D\times(0,T)$ where D is a bounded, simply connected domain whose boundary ∂D is three times continuously differentiable and let $u(x,y,t)$ be continuous in $\bar{D}\times[0,T]$. We assume that $c(x,y)$ and $d(x,y)$ are entire functions of their independent complex variables and that for $(x,y)\in\bar{D}$ we have $d(x,y) > 0$. By means of the change of variables $u = e^{\alpha t}v$ where $\alpha > 0$ is large, it is seen that without loss of generality we can assume $c(x,y) \leqslant 0$ for $(x,y)\in D$. From the maximum principle for parabolic equations and the Weierstrass approximation theorem we can assume without loss of generality (for purposes of approximation) that the boundary data assumed by $u(x,y,t)$ on $\partial D\times[0,T]$ is a polynomial in t, i.e.

$$u(x,y,t) = \sum_{n=0}^{N} f_n(x,y)t^n \quad ; \quad (x,y,t)\in\partial D\times[0,T] \qquad (3.2.19)$$

where (by a further approximation) the $f_n(x,y)$ are Hölder continuous functions defined on ∂D. We now look for a real valued solution of (3.2.15) in the form

$$w(x,y,t) = \sum_{n=0}^{N} w_n(x,y)t^n \tag{3.2.20}$$

such that $w(x,y,t) = u(x,y,t)$ for $(x,y,t)\varepsilon\partial D\times[0,T]$. From (3.2.15) and (3.2.19) it is seen that the function $w_n(x,y)$ must satisfy the recursive scheme

$$\begin{aligned}
&\frac{\partial^2 w_N}{\partial x^2} + \frac{\partial^2 w_N}{\partial y^2} + c(x,y)w_N = 0 \quad ; \quad (x,y)\varepsilon D \\
&w_N(x,y) = f_N(x,y) \quad ; \quad (x,y)\varepsilon\partial D \\
&\frac{\partial^2 w_n}{\partial x^2} + \frac{\partial^2 w_n}{\partial y^2} + c(x,y)w_n = (n+1)d(x,y)w_{n+1} \quad ; \quad (x,y)\varepsilon D \\
&w_n(x,y) = f_n(x,y) \quad ; \quad (x,y)\varepsilon\partial D,
\end{aligned} \tag{3.2.21}$$

for $n=0,1,\ldots,N-1$. The existence of the $w_n(x,y)$ for $n=0,1,\ldots,N$ follows from the smoothness of ∂D and the fact that $c(x,y) \leqslant 0$ in D (c.f. [27]). From Corollary 1.1.1, Theorem 1.3.3, Theorem 1.3.4 and the fact that $w_n(x,y)$ depends continuously on the nonhomogeneous term $(n+1)d(x,y)w_{n+1}(x,y)$, we can conclude that for $\varepsilon > 0$ there exists a real valued solution $w_1(x,y,t)$ of (3.2.15) which is an entire function of its independent complex variable such that

$$\max_{\bar{D}\times[0,T]} |w_1-w| < \varepsilon/2 \tag{3.2.22}$$

Now let $v(x,y,t) = u(x,y,t) - w(x,y,t)$ and let λ_n and $\phi_n(x,y)$ be the eigenvalues and eigenfunctions respectively that correspond to the eigenvalue problem

$$\begin{aligned}
&\phi_{xx} + \phi_{yy} + c(x,y)\phi + \lambda d(x,y)\phi = 0 \quad ; \quad (x,y)\varepsilon D \\
&\phi(x,y) = 0 \quad ; \quad (x,y)\varepsilon\partial D.
\end{aligned} \tag{3.2.23}$$

From (3.2.19)-(3.2.21) and the expansion theorem for the eigenvalue problem (3.2.23) (c.f. [35]) we can conclude that

$$v(x,y,t) = \sum_{n=0}^{\infty} a_n\phi_n(x,y)\exp(-\lambda_n t) \tag{3.2.24}$$

$$a_n = \iint_D v(x,y,0)\phi_n(x,y)d(x,y)dxdy$$

where the series in (3.2.24) converges absolutely and uniformly in $\bar{D}x[0,T]$. By truncating the series in (3.2.24) and appealing to Theorem 1.3.3 and Theorem 1.3.5 we can conclude that there exists a real valued solution $w_2(x,y,t)$ of (3.2.15) which is an entire function of its independent complex variables such that

$$\max_{\bar{D}x[0,T]} |w_2-v| < \varepsilon/2 \quad . \tag{3.2.25}$$

The inequalities (3.2.22) and (3.2.25) now imply that there exists a real valued solution $u_1(x,y,t)$ of (3.2.15) which is an entire function of its independent complex variables such that

$$\max_{\bar{D}x[0,T]} |u_1-u| < \varepsilon \quad . \tag{3.2.26}$$

Representing $u_1(x,y,t)$ in the form $u = \text{Re}\ \underset{\sim}{P}_2\{f\}$ and truncating the Taylor series for $f(z,t)$ to obtain the polynomial $f_n(z,t)$ such that for $(z,t)\varepsilon$ $\partial Dx[0,T]$, $|f-f_n|$ is sufficiently small, leads to the following theorem:

Theorem 3.2.3 ([13]): Let $u(x,y,t)$ be a real valued classical solution of (3.2.15) in $Dx(0,T)$ which ∂D is three times continously differentiable, $d(x,y) > 0$ in $\bar{D}$, and $u(x,y,t)$ is continuous in $\bar{D}x[0,T]$. Let $u_{nm}(x,y,t)$ for $n,m=0.1.2....$ be defined by (3.2.10). Then for any $\varepsilon > 0$ there exists integers $N=N(\varepsilon)$, $M=M(\varepsilon)$, and constants a_{nm}, $n=0,1,...,N$, $m=0,1,...,M$, such that

$$\max_{\bar{D}x[0,T]} \left|u- \sum_{n=0}^{N} \sum_{m=0}^{M} a_{nm}u_{nm}\right| < \varepsilon. \tag{3.2.27}$$

Theorem 3.2.3 can be used to provide a method for approximating solutions to initial-boundary value problems for (3.2.15) in the same manner as we have already done for elliptic equations in two independent variables and parabolic equations in one space variable. In particular we orthonormalize the set $\{u_{nm}(x,y,t)\}$ in the L^2 norm over the base and lateral boundary of the cylinder $D\times(0,T)$ to obtain the complete set $\{\phi_n(x,y,t)\}$. An approximate solution to the initial-boundary value problem on compact subsets of $D\times(0,T)$ is then given by

$$u^N(x,y,t) = \sum_{n=0}^{N} \sum_{m=0}^{M} c_{nm}\phi_n(x,y,t) \tag{3.2.28}$$

where

$$c_{nm} = \iint_{\partial D} u(x,y,t)\phi_{nm}(x,y,t)ds + \iint_{D} u(x,y,0)\phi_{nm}(x,y,0)dxdy \tag{3.2.29}$$

and ds denotes an element of surface area on ∂D.

Error estimates can again be found by applying the maximum principle. This procedure is particularly simple in the case of the heat equation where a complete family is given by (3.2.13) or (3.2.14)

3.3 The Inverse Stefan Problem.

In this section we will present an inverse method for constructing analytic solutions to the single phase Stefan problem for the heat equation in two space dimensions (For the case of one space dimension see the Introduction and section 2.3). Our solution of the inverse Stefan problem will be accomplished by assuming a priori that the free boundary is a relatively simple analytic surface and then constructing a solution to the heat equation which has this prescribed surface as a free boundary ([18]). Provided the solution is analytic in a sufficiently large domain we can then determine the initial-boundary data which is compatible with the given "free" boundary.

In physical terms we are asking the question of how must a given solid (.e.g. ice) be heated in order for it to melt in a prescribed manner, and by constructing a variety of such examples a qualitative idea can be obtained on the shape of the free boundary as a function of the initial-boundary conditions. As in the case of the inverse Stefan problem in one space dimension, such an inverse approach leads to two main problems. The first of these is that the inverse problem has its mathematical formulation as a non-characteristic Cauchy problem for the heat equation and is thus improperly posed in the real domain. However such a problem is well posed in the complex domain, and hence we are led to examine solutions of the heat equation in the space of several complex variables. The inverse Stefan problem can now locally be solved by appealing to the Cauchy-Kowalewski theorem (c.f. [21], [29]). However in addition to being far too tedious for practical computation and error estimation, such an approach does not provide us with the required global solution to the Cauchy problem under investigation. Hence we are led to the problem of the analytic continuation of solutions to non-characteristic Cauchy problems for the heat equation. We will accomplish this by using contour integration and the calculus of residues in the space of several complex variables to arrive at an explicit (global) series representation of the solution to the inverse Stefan problem.

We will motivate the mathematical formulation of the inverse Stefan problem in terms of an ice-water system undergoing a change of phase. Assume that a bounded simply connected region R with boundary ∂R is filled with ice at 0° Centigrade. Beginning at time $t=0$ a non-negative temperature $\gamma=\gamma(x,y,t)$ (where $\gamma(x,y,0)=0$) is applied to ∂R. The ice begins to melt and we will let the interphase boundary $\Gamma(t)$ between ice and water be described by $\Gamma(t) = \{(x,y) : \phi(x,y,t) = 0\}$ with the water lying in the region $\Phi(x,y,t) < 0$. The differential equation and boundary conditions governing the conduction of heat in the water are given by

$$\frac{\partial^2 u}{\partial x^2} + \frac{\partial^2 u}{\partial y^2} = \frac{1}{a}\frac{\partial u}{\partial t} \quad ; \quad \Phi(x,y,t) < 0 \tag{3.3.1}$$

$$u\Big|_{\partial R} = \gamma(x,y,t) \tag{3.3.2}$$

$$u\Big|_{\Gamma(t)} = 0 \quad , \quad k\frac{\partial u}{\partial \nu}\Big|_{\Gamma(t)} = \frac{\lambda\rho}{|\nabla\Phi|}\frac{\partial \Phi}{\partial t}\Big|_{\Gamma(t)} \tag{3.3.3}$$

where ν is the unit normal with respect to the space variables that points into the region $\Phi(x,y,t) < 0$, ∇ denotes the gradient with respect to the space variables, $u(x,y,t)$ is the temperature, a the diffusivity coefficient, λ the latent heat of fusion, ρ the density,and k the conductivity of the water. The Stefan problem is to find $\Gamma(t)$ and $u(x,y,t)$ given the function $\gamma(x,y,t)$. The inverse Stefan problem which we are interested in is to find $u(x,y,t)$ (an in particular $\gamma(x,y,t) = \lim_{(x,y)\to R} u(x,y,t)$) given $\Gamma(t)$. In general we cannot hope to solve the inverse problem for arbitrary $\Gamma(t)$; however by suitably restricting $\Gamma(t)$ to lie in a certain class of analytic surfaces we will be able to obtain a relatively simple series representation of the solution, and it is to this problem we now address ourselves.

Let D_t, $0 \leq t < t_o$, be a family of simply connected domains which depend analytically on a parameter t such that $\bigcup_{0\leq t<t_o} D_t$ contains $R \cup \partial R x[0,t_o]$. Let $z=\phi(\zeta,t)$ conformally map the unit disc Ω onto D_t (D_t being such that the image of $(-1,1)$ intersects R) and for $\zeta^*\varepsilon\Omega$, $0 \leq t < t_o$, define $\bar{\phi}(\zeta^*,t)$ by $\bar{\phi}(\zeta^*,t) = \overline{\phi(\bar{\zeta}^*,t)}$ where bars denote conjugation. Now set $z^* = \bar{\phi}(\zeta^*,t)$ and note that $z^*=\bar{z}$ if and only if $\zeta^*=\bar{\zeta}$. We now define the function $\Phi(x,y,t)$ for (possibly) complex values of x,y and t by

$$\Phi(x,y,t) = \frac{1}{2i}\left[\phi^{-1}(z,t) - \bar{\phi}^{-1}(z^*,t)\right] \tag{3.3.4}$$

where $z=x+iy$, $z^*=x-iy$. Noting that $z^*=\bar{z}$ if and only if x and y are real it is seen that $\Phi(x,y,t)=0$ corresponds to Im $\zeta=0$, i.e. the interval (-1,1) in the complex ζ plane. Similarly, the region $\Phi(x,y,t) < 0$ corresponds to Im $\zeta < 0$, i.e. the part of Ω which lies in the lower half plane. We finally note that for $z=x+iy\varepsilon\Gamma(t)$ we have $\phi^{-1}(z,t)=\bar{\phi}^{-1}(z^*,t)$ and hence

$$\left.\frac{\partial\Phi}{\partial t}\right|_{\Gamma(t)} = \frac{1}{2i}\left[\frac{\partial\phi^{-1}(z,t)}{\partial t} - \frac{\partial\bar{\phi}^{-1}(z^*,t)}{\partial t}\right]\Bigg|_{z^*=\bar{\phi}(\phi^{-1}(z,t),t)} \tag{3.3.5}$$

$$= g(z,t)$$

i.e. $\frac{\partial\Phi}{\partial t}$ restricted to $\Gamma(t)$ can be analytically continued (for each fixed t) to an analytic function of z for $z\varepsilon D_t$.

We will now construct a solution of (3.3.1) which has $\Phi(x,y,t)$ (as given by (3.3.4)) as a free boundary. In (3.3.1) we consider x and y as independent complex variables and define the transformation of $\mathbb{C}^2$ into itself by

$$\begin{aligned} z &= x+iy \\ z^* &= x-iy \quad . \end{aligned} \tag{3.3.6}$$

Under this transformation (3.3.1)-(3.3.3) become

$$L[U] \equiv \frac{\partial^2 U}{\partial z\partial z^*} - \frac{1}{4a}\frac{\partial U}{\partial t} = 0 \tag{3.3.7}$$

$$U(\phi(s,t),\bar{\phi}(s,t),t) = 0 \; ; \quad -1 < s < 1 \tag{3.3.8}$$

$$\begin{aligned} &U_1(\phi(s,t),\bar{\phi}(s,t),t)\,\frac{\partial\phi(s,t)}{\partial s} - U_2(\phi(s,t),\bar{\phi}(s,t),t)\,\frac{\partial\bar{\phi}(s,t)}{\partial s} \\ &= \frac{i\lambda\rho}{k}\left|\frac{\partial\phi(s,t)}{\partial s}\right|^2 g(\phi(s,t),t) \quad ; \quad -1 < s < 1 \end{aligned} \tag{3.3.9}$$

where $U(z,z^*,t) = u(\frac{z+z^*}{2}, \frac{z-z^*}{2i}, t)$, $g(z,t)$ is defined by (3.3.5), and subscripts denote differentiation of $U(z,z^*,t)$ with respect to the first and second variables respectively. (3.3.9) was arrived at in the following manner. Let $\zeta = \xi_1+i\xi_2$. Then C(t) is the image of $\xi_2=0$ under the mapping $z=\phi(\zeta,t)$. We have $\frac{\partial u}{\partial \nu} = \frac{\partial u}{\partial \xi_1}\frac{\partial \xi_1}{\partial \nu} + \frac{\partial u}{\partial \xi_2}\frac{\partial \xi_2}{\partial \nu}$. But on C(t), ν is in the direction of the level curve ξ_1 = constant, since $\zeta=\xi_1(x,y,t)+i\xi_2(x,y,t)=\phi^{-1}(z,t)$ is a conformal mapping. Hence on C(t) $\frac{\partial \xi_1}{\partial \nu} = 0$, and by the Cauchy Riemann equations and the fact that $\nu = -\frac{\nabla\xi_2}{|\nabla\xi_2|}$, we have $\frac{\partial \xi_2}{\partial \nu} = -\left|\frac{\partial \phi^{-1}(z,t)}{\partial z}\right|$ on C(t). Therefore $\frac{\partial u}{\partial \nu} = -\frac{\partial u}{\partial \xi_2}\left|\frac{\partial \phi^{-1}(z,t)}{\partial z}\right|$. But from $\phi^{-1}(\phi(z,t),t)=z$ we have

$$\frac{\partial \phi^{-1}}{\partial z}(\phi(z,t),t)\,\frac{\partial \phi(z,t)}{\partial z} = 1, \text{ i.e. } \frac{\partial \phi^{-1}(z,t)}{\partial z} = \frac{1}{\frac{\partial \phi}{\partial z}(\phi^{-1}(z,t),t)} .$$

Hence on C(t), $\frac{\partial \phi^{-1}(z,t)}{\partial z} = \frac{1}{\frac{\partial \phi}{\partial s}(s,t)}$, and therefore

$$\frac{\partial u}{\partial \nu} = -\frac{\partial u}{\partial \xi_2}\frac{1}{\left|\frac{\partial \phi}{\partial s}(s,t)\right|} = -i\left(U_1\frac{\partial \phi}{\partial s} - U_2\frac{\partial \bar{\phi}}{\partial s}\right)\left|\frac{\partial \phi}{\partial s}(s,t)\right|^{-1}.$$

(3.3.9) now follows from (3.3.3), (3.3.5), and the fact that $|\nabla\Phi|^2 = \left|\frac{\partial \phi^{-1}(z,t)}{\partial z}\right|\left|\frac{\partial \bar{\phi}^{-1}(\bar{z},t)}{\partial z}\right|$, which implies that on C(t) $|\nabla\Phi| = \frac{1}{\left|\frac{\partial \phi}{\partial s}(s,t)\right|}$.

Now let M be the adjoint operator defined by

$$M[V] \equiv \frac{\partial^2 V}{\partial z \partial z^*} + \frac{1}{4a}\frac{\partial V}{\partial t} = 0 \tag{3.3.10}$$

and let V be the fundamental solution of $M[V] = 0$ defined by

$$V(z,z^*,t\ ;\ \xi,\bar{\xi},\tau) = \frac{1}{t-\tau}\exp\left\{\frac{(z-\xi)(z^*-\bar{\xi})}{4a(t-\tau)}\right\} \tag{3.3.11}$$

where $\xi = \xi_1+i\xi_2$, $\bar{\xi} = \xi_1-i\xi_2$. Note that V satisfies the Goursat data

$$V(z,\bar{\xi},t\ ;\ \xi,\bar{\xi},\tau) = \frac{1}{t-\tau} \tag{3.3.12a}$$

$$V(\xi, z^*,t\ ;\ \xi,\bar{\xi},\tau) = \frac{1}{t-\tau} . \tag{3.3.12b}$$

We will now obtain the solution of the inverse Stefan problem (3.3.7)-(3.3.9) by first using Stokes theorem to integrate $VL[U] - UM[V]$ over a torus lying in the space of three complex variables and then computing the residue of the resulting integral representation.

Let τ be real and for t on the circle $|t-\tau|=\delta$, $\delta > 0$, let G(t) be a cell whose boundary consists of a curve C(t) lying on the surface $\phi^{-1}(z,t)=\bar{\phi}^{-1}(z^*,t)$ and line segments lying on the characteristic planes $z=\xi$ and $z^*=\bar{\xi}$ respectively which join the point $(\xi,\bar{\xi})$ to C(t). Now use Stokes theorem to integrate the identity

$$\begin{aligned} VL[U] - UM[V] &= (\tfrac{1}{2}VU_{z^*} - \tfrac{1}{2}V_{z^*}U)_z \\ &+ (\tfrac{1}{2}VU_z - \tfrac{1}{2}V_zU)_{z^*} \\ &- (\tfrac{1}{4a}\, VU)_t \end{aligned} \tag{3.3.13}$$

over the torus $\{(z,z^*,t) : (z,z^*)\varepsilon G(t), |t-\tau|=\delta\}$, making use of the initial conditions (3.3.12a), (3.3.12b) satisfied by V, the fact that U=0 on C(t), and the fact that $dzdz^*=0$ on $\partial G(t)x\Omega$ (complex differentials are interpreted in the sense of exterior differential forms c.f. [7]). After computing the residue at the point $z=\xi$, $z^*=\bar{\xi}$, this calculation gives ([18])

$$\begin{aligned} U(\xi,\bar{\xi},\tau) &= \frac{1}{4\pi i}\int_{|t-\tau|=\delta}\int_{C(t)} [VU_z dz - VU_{z^*}dz^*]dt \\ &= \frac{\lambda\rho}{4\pi k}\int_{|t-\tau|=\delta}\int_{\bar{\phi}^{-1}(\bar{\xi},t)}^{\phi^{-1}(\xi,t)} \frac{1}{t-\tau}\exp\{\frac{(\phi(\zeta,t)-\xi)(\bar{\phi}(\zeta,t)-\xi)}{4a(t-\tau)}\}\,. \qquad (3.3.14) \\ &\cdot |\frac{\partial\phi(\zeta,t)}{\partial\zeta}|^2 g(\phi(\zeta,t),t)d\zeta dt \\ &= \frac{i\lambda\rho}{2k}\sum_{n=0}^{\infty}\frac{1}{(4a)^n(n!)^2}\frac{\partial^n}{\partial\tau^n}\{\int_{\bar{\phi}^{-1}(\bar{\xi},\tau)}^{\phi^{-1}(\xi,\tau)} |\phi(\zeta,\tau)-\xi|^{2n}. \\ &\cdot |\frac{\partial\phi(\zeta,\tau)}{\partial\zeta}|^2 g(\phi(\zeta,\tau),\tau)d\zeta\} \end{aligned}$$

where $|\frac{\partial\phi}{\partial\zeta}(\zeta,\tau)|^2 = \frac{\partial\phi}{\partial\zeta}(\zeta,\tau)\,\frac{\partial\bar{\phi}}{\partial\zeta}(\zeta,\tau)$ and $|\phi(\zeta,\tau) - \xi|^2 = (\phi(\zeta,\tau)-\xi)(\bar{\phi}(\zeta,\tau)-\bar{\xi})$. The equation (3.3.14) is the solution of the inverse Stefan problem, i.e. for every one parameter family of conformal mappings $\phi(z,t)$, (3.3.14) defines a solution of (3.3.1)-(3.3.3) with the "free" boundary $\Gamma(t)=\{(x,y):\Phi(x,y,t)=0\}$ given by (3.3.4). Note that from the definition of the conformal mappings $\phi(z,t)$, it is seen that (3.3.14) is valid in a region containing $R \cup \partial R\times[0,t_o]$. In order to obtain a physically meaningful solution of the inverse Stefan problem we assume $\gamma(x,y,t)=0$ for $(x,y,t)\varepsilon\partial R\times[0,t_o] \cap \{(x,y,t):\Phi(x,y,t) \geqslant 0\}$ and choose the conformal mappings $\phi(z,t)$ such that $u(x,y,t) \geqslant 0$ for $\{(x,y,t):\Phi(x,y,t) < 0\}$. We note that from the boundary condition (3.3.3) this last condition is always satisfied (at least for t_o sufficiently small) provided we choose $\phi(z,t)$ such that $\left.\frac{\partial\phi}{\partial t}\right|_{\Gamma(t)} \geqslant 0$. Due to the appearance of the factor $(n!)^2$ in the denominator of the nth term in the series (3.3.14), accurate approximation of the solution to the inverse Stefan problem can be obtained by truncating this series after only a few terms.

IV The method of ascent for elliptic equations

4.1 Interior Domains.

Although it is possible to extend some of the results of the last three chapters to partial differential equations in three and four independent variables (c.f. [8],[31],[48]) the analysis becomes increasingly more complicated, and hence somewhat less practical for purposes of analytic and numerical approximation. However in certain special cases it is possible to make such an extension in a rather simple and straightforward fashion, and it is this topic which we will consider in this chapter. The special case we have in mind is the elliptic equation

$$\Delta_n u + B(r^2)u = 0 \tag{4.1.1}$$

where $B(r^2)$ is a real valued entire function of $r^2 = x_1^2 + \ldots + x_n^2$, and we will first consider solutions of (4.1.1) which are defined in interior domains. The theory of (4.1.1) in interior domains was developed by R.P. Gilbert in [32], (who described his theory as a "method of ascent") and for exterior domains by Colton and Wendland in [20]. Extensions of this development to the case of parabolic and pseudoparabolic equations are also possible (c.f. [46]), although we will not discuss this topic in the present work.

Equations in the form (4.1.1) arise naturally in the theory of steady state heat conduction and the scattering of acoustic waves (to name but two areas of many possible applications) when the medium is no longer homogeneous but varies smoothly as a function of the variable r (c.f. [6], [20]).

We begin our study of (4.1.1) in interior domains by first considering the case n=2 and using the Bergman operator (section 1.3) to represent real value twice continuously differentiable solutions of (4.1.1) in the form

$$u(x,y) = \operatorname{Re} \int_{-1}^{1} E(z,\bar{z},t) f\left(\frac{z}{2}(1-t^2)\right) \frac{dt}{\sqrt{1-t^2}} \tag{4.1.2}$$

where $z = x+iy$, $\bar{z} = x-iy$, and

$$E(z,z^*,t) = 1 + \sum_{k=1}^{\infty} t^{2k} z^k \int_0^{z^*} P^{(2k)}(z,z^*)\,dz^* \tag{4.1.3}$$

with the $P^{(2k)}$ defined recursively by

$$\begin{aligned} P^{(2)} &= -2B \\ (2k+1)P^{(2k+2)} &= -2\left[P_z^{(2k)} + B \int_0^{z^*} P^{(2k)}\,dz^*\right] \quad ; \quad k \geq 1 . \end{aligned} \tag{4.1.4}$$

Recall from section 1.3 that $f(z)$ is an analytic function of z in some neighbourhood of the origin. At this point we make the assumption that $u(x,y)$ is defined in the interior of a bounded domain D containing the origin where D is strictly starlike with respect to the origin, i.e. if P is a point in $\bar{D} = D \cup \partial D$, then the line segment $\overline{OP}$ is contained in D except for possibly the endpoint P. We will further assume that ∂D is twice continuously differentiable. Throughout this chapter whenever we refer to a domain D we will assume it satisfies the conditions described above.

Returning now to (4.1.2)-(4.1.4) we have the following lemma.

<u>Lemma 4.1</u> ([2]): For (4.1.1) the generating function $E(z,z^*,t)$ is a real valued entire function of $r^2 = zz^*$ and t, i.e. $E(z,z^*,t) = \tilde{E}(r^2,t)$.

<u>Proof</u>: Let

$$Q^{(2k)}(z,z^*) = z^k \int_0^{z^*} P^{(2k)}(z,z^*)\,dz^* \quad ; \quad k = 1,2 \ldots \tag{4.1.5}$$

Then from (4.1.4) we have

$$Q_{z^*}^{(2)} + 2zB(r^2) = 0 \tag{4.1.6a}$$

$$(2k+1)Q_{z^*}^{(2k+2)} + 2z\left[Q_{zz^*}^{(2k)} + B(r^2)Q^{(2k)} - \frac{k}{z} Q_{z^*}^{(2k)}\right] = 0 \tag{4.1.6b}$$

$$Q^{(2k)}(z,0) = 0 \qquad (4.1.6c)$$

for k=1,2,... . (4.1.6a) can be rewritten in the form

$$\frac{\partial Q^{(2)}}{\partial (r^2)} + 2B(r^2) = 0, \qquad (4.1.7a)$$

and if we require $Q^{(2)}(0) = 0$ it is seen that $Q^{(2)}$ depends only on r^2 and satisfies (4.1.6c). Now assume that $Q^{(2k)}$ depends only on r^2. Then (4.1.6b) and (4.1.6c) will be satisfied if $Q^{(2k+2)}$ is a solution of

$$(2k+1)\frac{\partial Q^{(2k+2)}}{\partial (r^2)} + 2\left[\frac{\partial\left(r^2 \frac{\partial Q^{(2k)}}{\partial (r^2)}\right)}{\partial (r^2)} + B(r^2)Q^{(2k)} - k\frac{\partial Q^{(2k)}}{\partial (r^2)}\right] = 0 \qquad (4.1.7b)$$

such that $Q^{(2k+2)}(0) = 0$. From (4.1.7b) we see that $Q^{(2k+2)}$ is a function only of r^2, and the lemma now follows by induction. The fact that $\tilde{E}(r^2,t)$ is entire follows from the fact that $E(z,z^*,t)$ is entire (section 1.2).

From lemma 4.1 we can write (4.1.2) in the form

$$u(x,y) = \int_{-1}^{1} \tilde{E}(r^2,t)H(x(1-t^2),y(1-t^2))\frac{dt}{\sqrt{1-t^2}} \qquad (4.1.8)$$

where

$$H(x,y) = \text{Re } f\left(\frac{z}{2}\right) \qquad (4.1.9)$$

is a harmonic function. From section 1.3 and lemma 4.1 it can be shown that $\tilde{E}(r^2,t)$ satisfies the partial differential equation

$$(1-t^2)\tilde{E}_{rt} - \frac{1}{t}\tilde{E}_r + rt\left[\tilde{E}_{rr} + \frac{1}{r}\tilde{E}_r + B\tilde{E}\right] = 0, \qquad (4.1.10)$$

the initial condition

$$\tilde{E}(0,t) = 1, \qquad (4.1.11)$$

and has a series expansion of the form

$$\tilde{E}(r^2,t) = 1 + \sum_{k=1}^{\infty} t^{2k}e^{(k)}(r^2) \qquad (4.1.12)$$

which converges absolutely and uniformly for t and r arbitrarily large (but bounded). Now define the harmonic function $h(x,y)$ by

$$h(x,y) = \int_{-1}^{1} H(x(1-t^2),\ y(1-t^2)) \frac{dt}{\sqrt{1-t^2}} \ . \tag{4.1.13}$$

Then (4.1.8) can be rewritten as

$$u(x,y) = h(x,y) + \int_0^1 \sigma G(r^2,1-\sigma^2)h(x\sigma^2,y\sigma^2)d\sigma \tag{4.1.14}$$

where

$$G(r^2,\rho) = \sum_{k=1}^{\infty} \frac{2e^{(k)}(r^2)\Gamma(k+\frac{1}{2})}{\Gamma(\frac{1}{2})\Gamma(k)} \rho^{k-1} \ . \tag{4.1.15}$$

These last two equations follow immediately from expanding $h(x,y)$ and $H(x,y)$ in a series of harmonic polynomials, integrating (4.1.8) termwise using (4.1.12), and using the elementary properties of the Beta function. From section 1.3 it is clear that (4.1.14) defines a mapping of the class of real valued harmonic functions defined in D onto the class of real valued solutions of (4.1.1) (for n=2) defined in D.

We now want to generalize the representation (4.1.14) from n=2 to general n. To this end we first look for real valued twice continuously differentiable solutions of (4.1.1) in the form

$$u(\underset{\sim}{x}) = \int_0^1 t^{n-2}E(r^2,t;n)H(\underset{\sim}{x}(1-t^2)) \frac{dt}{\sqrt{1-t^2}} \tag{4.1.16}$$

where $\underset{\sim}{x} = (x_1,\dots,x_n)$ and $H(\underset{\sim}{x})$ is a real valued harmonic function in D (which is now of course a domain in $\mathbb{R}^n$). We require that $E(r^2,t;n)$ be an entire function of t and r^2 and satisfy the initial condition $E(0,t;n)=1$. We now temporarily replace the path of integration from zero to one by a loop starting from s=+1, passing counterclockwise around the origin and onto the second sheet of the Riemann surface of the integrand, and then back up to t=+1, and substitute the resulting expression into the differential equation (4.1.1). If $u(\underset{\sim}{x})$ is to be a solution of (4.1.1), it is then easily verified by integrating by parts that $E(r^2,t;n)$ must satisfy the singular partial differential equation

$$(1-t^2)E_{rt} + \frac{n-3}{t} E_r + rs\left[E_{rr} + \frac{1}{r} E_r + BE\right] = 0 \quad . \tag{4.1.17}$$

We now look for a solution of (4.1.17) in the form

$$E(r^2,t;n) = 1 + \sum_{k=1}^{\infty} t^{2k} e^{(k)}(r^2;n) \quad . \tag{4.1.18}$$

Substituting (4.1.18) into (4.1.17) yields the following recursion formulas for the determination of the $e^{(k)}(r^2;n)$:

$$\begin{aligned} &(n-1)e_r^{(1)} = -rB \\ &(2k+n-3)e_r^{(k)} = (2k-3)e_r^{(k-1)} - re_{rr}^{(k-1)} - rBe^{(k-1)}; \\ &\qquad\qquad\qquad\qquad k \geqslant 2. \end{aligned} \tag{4.1.19}$$

From the initial condition $E(0,t;n) = 0$ we have the initial conditions

$$e^{(k)}(0;n) = 0 \quad ; \quad k = 1,2,\ldots \quad . \tag{4.1.20}$$

Hence, each of the $e^{(k)}(r^2;n)$ in (4.1.18) is uniquely determined. We must now show that the series (4.1.18) converges for t and r arbitrarily large (but bounded). We first note that for $n=2$ the $e^{(k)}(r^2;2)$ are identical with the functions $e^{(k)}(r^2)$ defined by (4.1.12). This follows from the facts that the form of the series expansion for $\tilde{E}(r^2,t)$ and $E(r^2,t;2)$ are the same and these functions satisfy the same differential equation and initial condition. Hence, the series (4.1.18) converges when $n=2$. Now define functions $c^{(k)}(r^2;n)$ by the formula

$$c^{(k)}(r^2;n) = \frac{2e^{(k)}(r^2;n)\Gamma(k+\frac{n}{2}-\frac{1}{2})}{\Gamma(\frac{n}{2}-\frac{1}{2})\Gamma(k)} \quad ; \quad k \geqslant 1 \quad . \tag{4.1.21}$$

Then from (4.1.19) and (4.1.20) it is seen that the $c^{(k)}(r^2;n)$ satisfy the recursion formula

$$c_r^{(1)} = -rB \qquad (4.1.22)$$

$$2(k-1)c_r^{(k)} = (2k-3)c_r^{(k-1)} - rc_{rr}^{(k-1)} - rBc^{(k-1)} \quad ; \quad k \geqslant 2$$

and the initial conditions

$$c^{(k)}(0;n) = 0 \quad ; \quad k \geqslant 1 \ . \qquad (4.1.23)$$

(4.1.22) and (4.1.23) imply that the $c^{(k)}(r^2;n)$ are in fact independent of n. Since we know the series (4.1.18) is convergent when n=2, we can now conclude from (4.1.21) and the fact that the $c^{(k)}(r^2;n)$ are independent of n that the series (4.1.18) converges absolutely and uniformly for r and t arbitrarily large (but bounded). This establishes the existence of the operator defined by (4.1.16) and (4.1.18). If in this operator we now set

$$h(\underset{\sim}{x}) = \int_0^1 t^{n-2} H(\underset{\sim}{x}(1-t^2)) \frac{dt}{\sqrt{1-t^2}} \qquad (4.1.24)$$

we arrive at the following integral operator which maps real valued harmonic functions defined in D into the class of real valued solutions of (4.1.1) defined in D:

$$u(\underset{\sim}{x}) = (\underset{\sim}{I}+\underset{\sim}{G})h = h(\underset{\sim}{x}) + \int_0^1 \sigma^{n-1} G(r^2, 1-\sigma^2) h(\underset{\sim}{x}, \sigma^2) d\sigma \qquad (4.1.25)$$

where $G(r^2,\rho)$ is defined by (4.1.15) and is independent of n. This last fact is the basis for referring to the approach used in this section as a "method of ascent".

We now want to show that the operator $\underset{\sim}{I}+\underset{\sim}{G}$ is invertible, i.e. for every solution $u(\underset{\sim}{x})$ of (4.1.1) in D there exists a harmonic function $h(\underset{\sim}{x})$ in D such that (4.1.25) is valid. To this end we rewrite (4.1.25) as the Volterra integral equation

$$\Phi(r;\theta;\phi) = \psi(r;\theta;\phi) + \int_0^r K(r,\rho)\psi(\rho;\theta;\phi) d\rho \qquad (4.1.26)$$

where

$$\Phi(r;\theta;\phi) = r^{(n-2)/2}u(r;\theta;\phi)$$

$$\psi(r;\theta;\phi) = r^{(n-2)/2}h(r;\theta;\phi) \qquad (4.1.27)$$

$$K(r,\rho) = \frac{1}{2r}\, G(r^2,\ 1-(\ \frac{\rho}{r}\ ))$$

and $(r;\theta;\phi)$ are spherical coordinates. From the recursion formula (4.1.22) it is seen that each $c^{(k)}(r^2;n)$ is of the form

$$c^{(k)}(r^2;n) = r^{2k}\tilde{c}(r^2;n) \qquad (4.1.28)$$

where $\tilde{c}^{(k)}(r^2;n)$ is an entire function of r^2. This follows from the fact that the differential operator $(2k-3)(\frac{d}{dr}) - r(\frac{d^2}{dr^2})$ annihilates r^{2k-2}.

Hence the function $K(r,\rho)$ defined in (4.1.27) is an entire function of r and ρ. Since (4.1.26) is a Volterra integral equation of the second kind it is now clear that there exists a unique solution $\psi(r;\theta;\phi)$ of (4.1.26). From the fact that

$$0 = \Delta_n u+B(r^2)u = \Delta h + \int_0^1 \sigma^{n-1}G(r^2,1-\sigma^2)\Delta h(\underset{\sim}{x}\sigma^2)d\sigma \qquad (4.1.29)$$

it can easily be seen that $r^{-(n-2)/2}\psi(r;\theta;\phi) = h(r,\theta;\phi)$ is a harmonic function in D (rewrite (4.1.29) in the form (4.1.26) where Φ now equals zero, and appeal to the uniqueness of solutions to Volterra integral equations of the second kind). We can now conclude that the operator $\underset{\sim}{I}+\underset{\sim}{G}$ is invertible.

We summarize our results in the following theorem:

<u>Theorem 4.1.1</u> ([32]): Let $u(\underset{\sim}{x})$ be a real valued twice continuously differentiable solution of (4.1.1) in D where D is strictly starlike with respect to the origin. Then $u(\underset{\sim}{x})$ can be represented in the form $u(\underset{\sim}{x})=(\underset{\sim}{I}+\underset{\sim}{G})h$ where $h(\underset{\sim}{x})$ is a real valued harmonic function in D. Conversely, if $h(\underset{\sim}{x})$ is harmonic in D, then $u(\underset{\sim}{x}) = (\underset{\sim}{I}+\underset{\sim}{G})h$ is a solution of (4.1.1) in D.

Remark 1: The assumption that $B(r^2)$ is an entire function can be considerably weakened. This follows from the fact that it can be shown ([31]) that

$$G(r,1-\sigma^2) = -2r\, R_3(r,r;r\sigma^2,0) \tag{4.1.30}$$

where $R(x,y;\xi,\eta)$ is the Riemann function for the hyperbolic equation

$$u_{xy} + B(xy)u = 0 \quad , \tag{4.1.31}$$

and the subscript denotes differentiation with respect to ξ. Hence if $\tilde{B}(r) = B(r^2)$ is continuously differentiable we can conclude that $G(r,1-\sigma^2)$ exists and is twice continuously differentiable.

Remark 2: An alternate approach to constructing the operator $\underset{\sim}{I}+\underset{\sim}{G}$ has been outlined by M. Eichler in [23]. This approach is somewhat similar to that which we will use in section 4.2 to obtain a "method of ascent" for equations of the form (4.1.1) definied in exterior domains.

We will now show how the integral operator $\underset{\sim}{I}+\underset{\sim}{G}$ can be used to solve (interior) boundary value problems for (4.1.1). To be specific we will consider the interior Dirichlet problem for (4.1.1) in the case n=3 under the assumption that $B(r^2) \leq 0$ in D; the same approach can be used to treat the Dirichlet, Neumann, and Robin problems for $n \geq 2$. We want to construct a solution $u(\underset{\sim}{x}) \in C^2(D) \cap C^0(\bar{D})$ of (4.1.1) in D such that $u(\underset{\sim}{x}) = f(\underset{\sim}{x})$ on ∂D where $f(\underset{\sim}{x})$ is a known continuous function defined on ∂D. We look for a solution in the form

$$u(\underset{\sim}{x}) = (\underset{\sim}{I}+\underset{\sim}{G})h \tag{4.1.32}$$

where $h(\underset{\sim}{x})$ is represented in terms of the double layer potential

$$h(\underset{\sim}{x}) = \frac{1}{2\pi} \int_{\partial D} \psi(\underset{\sim}{\xi}) \frac{\partial}{\partial \nu} \left(\frac{1}{R} \right) d\omega_{\underset{\sim}{\xi}} .$$

(This is similar to the approach used in section 1.3 for elliptic equations in two independent variables). In (4.1.33) $d\omega_{\underset{\sim}{\xi}}$ denotes an element of surface area on ∂D, $R = |\underset{\sim}{x} - \underset{\sim}{\xi}|$, ν is the inward normal to ∂D at the point $\underset{\sim}{\xi}$, and ψ is a continuous density to be determined. Substituting (4.1.33) into (4.1.) and interchanging the orders of integration gives

$$u(\underset{\sim}{x}) = \frac{1}{2\pi} \int_{\partial D} \psi(\underset{\sim}{\xi}) \frac{\partial}{\partial \nu} \left(\frac{1}{R} \right) d\omega_{\underset{\sim}{\xi}}$$

$$+ \frac{1}{2\pi} \int_{\partial D} \psi(\underset{\sim}{\xi}) \left\{ \int_0^1 \sigma^2 G(r^2, 1-\sigma^2) \frac{\partial}{\partial \nu} \left(\frac{1}{|\underset{\sim}{x}\sigma^2 - \underset{\sim}{\xi}|} \right) d\sigma \right\} d\omega_{\underset{\sim}{\xi}} \quad . \qquad (4.1.34)$$

We will show shortly that for $\underset{\sim}{\xi}$, $\underset{\sim}{x}$ on ∂D

$$\left| \int_0^1 \sigma^2 G(r^2, 1-\sigma^2) \frac{\partial}{\partial \nu} \left(\frac{1}{|\underset{\sim}{x}\sigma^2 - \underset{\sim}{\xi}|} \right) d\sigma \right| \leqslant \frac{\text{constant}}{|\underset{\sim}{x} - \underset{\sim}{\xi}|} \quad . \qquad (4.1.35)$$

Assuming this fact for the time being, we let $\underset{\sim}{x}$ tend to ∂D, and, using the discontinuity properties of double and single layer potentials (c.f.[21], [29]), we arrive at the following integral equation for $\psi(\underset{\sim}{\xi})$:

$$f(\underset{\sim}{x}) = \psi(\underset{\sim}{x}) + \frac{1}{2\pi} \int_{\partial D} \psi(\underset{\sim}{\xi}) \frac{\partial}{\partial \nu} \left(\frac{1}{R} \right) d\omega_{\underset{\sim}{\xi}}$$

$$\qquad (4.1.36)$$

$$+ \frac{1}{2\pi} \int_{\partial D} \psi(\underset{\sim}{\xi}) \left\{ \int_0^1 \sigma^2 G(r^2, 1-\sigma^2) \frac{\partial}{\partial \nu} \left(\frac{1}{|\underset{\sim}{x}\sigma^2 - \underset{\sim}{\xi}|} \right) d\sigma \right\} d\omega_{\underset{\sim}{\xi}}$$

$$= (\underset{\sim}{I} + \underset{\sim}{T})\psi \qquad ; \qquad \underset{\sim}{x} \in \partial D \quad .$$

Before discussing the invertibility of the operator $\underset{\sim}{I} + \underset{\sim}{T}$ we prove the extimate (4.1.35). Since $G(r^2, 1-\sigma^2)$ is continuous, there exists a positive constant C such that for $\underset{\sim}{\xi}$, $\underset{\sim}{x}$ on ∂D

$$\left| \int_0^1 \sigma^2 G(r^2, 1-\sigma^2) \frac{\partial}{\partial \nu} \left(\frac{1}{|\underset{\sim}{x}\sigma^2 - \underset{\sim}{\xi}|} \right) d\sigma \right|$$

$$\qquad (4.1.37)$$

$$\leqslant C \int_0^1 \frac{1}{|\underset{\sim}{x}\rho - \underset{\sim}{\xi}|^2} d\rho \quad .$$

We now examine the function $|\underset{\sim}{x}\rho-\underset{\sim}{\xi}|^{-2}$ for $\underset{\sim}{x}$ and $\underset{\sim}{\xi}$ on ∂D. Without loss of generality we can restrict our attention to values of $\underset{\sim}{x}$ and $\underset{\sim}{\xi}$ such that $\underset{\sim}{x}.\underset{\sim}{\xi} \geq 0$. This follows from the fact that if $\underset{\sim}{x}.\underset{\sim}{\xi} < 0$ then

$$\begin{aligned} |\underset{\sim}{x}\rho-\underset{\sim}{\xi}|^2 &= \rho^2|\underset{\sim}{x}|^2+|\underset{\sim}{\xi}|^2 - 2\rho\underset{\sim}{x}.\underset{\sim}{\xi} \\ &\geq \rho^2|\underset{\sim}{x}|^2+|\underset{\sim}{\xi}|^2 , \end{aligned} \tag{4.1.38}$$

and hence for such values of $\underset{\sim}{x}$ and $\underset{\sim}{\xi}$ the integral on the right hand side of (4.1.37) can be bounded by a constant independent of $\underset{\sim}{x}$ and $\underset{\sim}{\xi}$ (since D contains the origin). This in turn implies that (4.1.35) is valid. Hence we now assume that $\underset{\sim}{x}.\underset{\sim}{\xi} \geq 0$ and observe that either

$$|\underset{\sim}{x}\rho-\underset{\sim}{\xi}| \geq |\underset{\sim}{x}-\underset{\sim}{\xi}| \tag{4.1.39}$$

for $0 \leq \rho \leq 1$, or

$$|\underset{\sim}{x}\rho-\underset{\sim}{\xi}| \geq |\underset{\sim}{\xi}| \tag{4.1.40}$$

for $0 \leq \rho \leq 1$, or there exists a ρ_o, $0 < \rho_o < 1$, such that

$$|\underset{\sim}{x}\rho-\underset{\sim}{\xi}| \geq |\underset{\sim}{x}\rho_o-\underset{\sim}{\xi}| \tag{4.1.41}$$

for $0 \leq \rho \leq 1$, where

$$(\underset{\sim}{x}\rho_o-\underset{\sim}{\xi}).\underset{\sim}{x} = 0 \quad . \tag{4.1.42}$$

In the first two cases we can immediately conclude from (4.1.35) that an estimate of the form (4.1.37) is valid. Hence we now consider the third case. From (4.1.42) we have that $\rho_o|\underset{\sim}{x}|^2 = \underset{\sim}{\xi}.\underset{\sim}{x}$ and hence

$$\begin{aligned} \underset{\sim}{x}\rho_o-\underset{\sim}{\xi} &= \underset{\sim}{x}\frac{\underset{\sim}{\xi}.\underset{\sim}{x}}{|\underset{\sim}{x}|^2} - \underset{\sim}{\xi} \\ &= \underset{\sim}{x}-\underset{\sim}{\xi}-\underset{\sim}{x}\frac{(\underset{\sim}{x}-\underset{\sim}{\xi}).\underset{\sim}{x}}{|\underset{\sim}{x}|^2} \quad . \end{aligned} \tag{4.1.43}$$

Therefore

$$|\underset{\sim}{x}\rho_o-\underset{\sim}{\xi}|^2 = |\underset{\sim}{x}-\underset{\sim}{\xi}|^2 - \frac{((\underset{\sim}{x}-\underset{\sim}{\xi}).\underset{\sim}{x})^2}{|\underset{\sim}{x}|^2} \quad . \tag{4.1.44}$$

Since D is strictly starlike, there exists a positive constant $\alpha < 1$ which is independent of $\underset{\sim}{x}$ and $\underset{\sim}{\xi}$ such that

$$|(\underset{\sim}{x}-\underset{\sim}{\xi})\cdot\underset{\sim}{x}| \leqslant \alpha|\underset{\sim}{x}-\underset{\sim}{\xi}||\underset{\sim}{x}| \tag{4.1.45}$$

uniformly for all $\underset{\sim}{x},\underset{\sim}{\xi}\varepsilon\partial D$ such that $\underset{\sim}{x}.\underset{\sim}{\xi} \geqslant 0$. Hence from (4.1.41) we have

$$|\underset{\sim}{x}\rho_o-\underset{\sim}{\xi}|^2 \geqslant (1-\alpha^2)|\underset{\sim}{x}-\underset{\sim}{\xi}|^2 \tag{4.1.46}$$

uniformly for all $\underset{\sim}{x},\underset{\sim}{\xi}\varepsilon\partial D$ such that $\underset{\sim}{x}.\underset{\sim}{\xi} \geqslant 0$ and from (4.1.41)

$$\begin{aligned} |\underset{\sim}{x}\rho-\underset{\sim}{\xi}|^2 &= |\underset{\sim}{x}\rho_o-\underset{\sim}{\xi}|^2 + (\rho-\rho_o)^2 \\ &\geqslant (1-\alpha^2)|\underset{\sim}{x}-\underset{\sim}{\xi}|^2+(\rho-\rho_o)^2 . \end{aligned} \tag{4.1.47}$$

Therefore for $\underset{\sim}{x}.\underset{\sim}{\xi} \geqslant 0$ and the case (4.1.41) we have

$$\begin{aligned} \int_o^1 |\underset{\sim}{x}\rho-\underset{\sim}{\xi}|^{-2}d\rho &\leqslant \frac{1}{(1-\alpha^2)}\int_o^1 \frac{d\rho}{|\underset{\sim}{x}-\underset{\sim}{\xi}|^2+(\rho-\rho_o)^2} \\ &= \frac{1}{(1-\alpha^2)|\underset{\sim}{x}-\underset{\sim}{\xi}|} \arctan\left(\frac{\rho-\rho_o}{|\underset{\sim}{x}-\underset{\sim}{\xi}|}\right)\Bigg|_{\rho=0}^{\rho=1} \\ &\leqslant \frac{4\pi}{(1-\alpha^2)|\underset{\sim}{x}-\underset{\sim}{\xi}|} , \end{aligned} \tag{4.1.48}$$

and from a consideration of the remaining (trivial) cases we can now conclude from (4.1.37) that (4.1.35) is valid.

To complete our discussion of the Dirichlet problem for (4.1.1) we now show that $(\underset{\sim}{I}+\underset{\sim}{T})^{-1}$ exists where $\underset{\sim}{T}$ is defined in (4.1.36). Since we have already shown that $\underset{\sim}{T}$ has a weakly singular kernel, if $(\underset{\sim}{I}+\underset{\sim}{T})^{-1}$ exists then there are a variety of constructive methods for obtaining the unknown density $\psi(\underset{\sim}{\xi})$ and hence the solution of the Dirichlet problem for (4.1.1) (c.f. [1],[22]).

To show that $(\underset{\sim}{I}+\underset{\sim}{T})^{-1}$ exists, by the Fredholm alternative it suffices to show that if $(\underset{\sim}{I}+\underset{\sim}{T})\psi = 0$ then $\psi=0$. Suppose $(\underset{\sim}{I}+\underset{\sim}{T})\psi = 0$. Then the potential defined by (4.1.33) generates by (4.1.31)a solution $u(\underset{\sim}{x})$ of (4.1.1) such that $u(\underset{\sim}{x})=0$ for $\underset{\sim}{x}\varepsilon\partial D$. Since $B(r^2) \leqslant 0$ for $\underset{\sim}{x}\varepsilon D$, we can conclude from the maximum principle for elliptic equations that $u(\underset{\sim}{x})\equiv 0$ in D. This implies from our previous discussion that $h(\underset{\sim}{x})$ as defined by (4.1.33) is identically zero in D, and hence letting $\underset{\sim}{x}$ tend to ∂D and using the discontinuity properties of double layer potentials we have

$$0=\psi(\underset{\sim}{x})+\frac{1}{2\pi}\int_{\partial D}\psi(\underset{\sim}{\xi})\frac{\partial}{\partial\nu}\left(\frac{1}{R}\right)d\omega_{\xi}\ ;\ \underset{\sim}{x}\in\partial D \qquad (4.1.49)$$

But from classical results in potential theory (c.f. [29]) (4.1.49) implies that $\psi(\underset{\sim}{\xi})=0$ for $\underset{\sim}{\xi}\in\partial D$ and hence $(\underset{\sim}{I}+\underset{\sim}{T})^{-1}$ exists.

4.2 Exterior Domains.

We now want to obtain a "method of ascent" for solutions of equations of the form (4.1.1) which are defined in exterior domains. In the case when $B(r^2)$ decays sufficiently rapidly at infinity, a "method of ascent" for equations defined in exterior domains can be obtained by simply applying a Kelvin transformation (c.f. [29]) and then using the results of section 4.1. However if we are interested in problems which arise from scattering theory, then $B(r^2)$ does not tend to zero as r tends to infinity, and the above approach can no longer be used. It is this type of problem which we will be interested in for the remainder of this chapter. In particular the mathematical problems which we will consider in the present section have their origin in the following problem connected with the scattering of acoustic waves in a non homogeneous medium. Let an incoming plane acoustic wave of frequency ω moving in the direction of the z axis be scattered off a bounded rigid obstacle D which is surrounded by a pocket of rarefied or condensed air in which the local speed of sound is given by c(r) where $r=|\underset{\sim}{x}|$ for $\underset{\sim}{x}\in\mathbb{R}^3$. Assume that this pocket of air is contained in a ball of radius a and that for $r\geq a$ we have $c(r)=c_o=\text{constant}$. Let $U(\underset{\sim}{x})$ be the velocity potential (factoring out a term of the form $e^{i\omega t}$) and set $B(r)=\left(\frac{c_o}{c(r)}\right)^2-1$.

Then, assuming $|\nabla c(r)|$ is small compared with $\lambda c(r)$ where $\lambda = \frac{\omega}{c_o}$, we are led to the following boundary value problem, where $u_s(\underset{\sim}{x})$ is the velocity potential of the scattered wave and ν denotes the outward normal to ∂D:

$$U(\underset{\sim}{x}) = e^{i\lambda z} + u_s(\underset{\sim}{x}) \tag{4.2.1}$$

$$\Delta_3 U + \lambda^2(1+B(r))U = 0 \text{ in } \mathbb{R}^3\backslash D \tag{4.2.2}$$

$$\frac{\partial U}{\partial \nu} = 0 \text{ on } \partial D \tag{4.2.3}$$

$$\lim_{r\to\infty} r\left(\frac{\partial u_s}{\partial r} - i\lambda u_s\right) = 0 \tag{4.2.4}$$

where the Sommerfeld radiation condition (4.2.4) is assumed to hold uniformly in all directions. Now let

$$u_s(\underset{\sim}{x}) = v(\underset{\sim}{x}) + u(\underset{\sim}{x}) \tag{4.2.5}$$

where $v(\underset{\sim}{x}) \varepsilon C^2(\mathbb{R}^3\backslash\bar{D}) \cap C^1(\mathbb{R}^3\backslash D)$ is such that $e^{i\lambda z} + v(\underset{\sim}{x})$ is a solution of (4.2.2) in $\mathbb{R}^3\backslash\bar{D}$ and $v(\underset{\sim}{x})$ satisfies (4.2.4). If such a function $v(\underset{\sim}{x})$ can be found, then the boundary value problem (4.2.1)-(4.2.4) for $U(\underset{\sim}{x})$ can be reduced to the following boundary value problem for $u(\underset{\sim}{x})$:

$$\Delta_3 u + \lambda^2(1+B(r))u = 0 \text{ in } \mathbb{R}^3\backslash\bar{D} \tag{4.2.6}$$

$$\frac{\partial u}{\partial \nu} = f(\underset{\sim}{x}) \text{ on } \partial D \tag{4.2.7}$$

$$\lim_{r\to\infty} r\left(\frac{\partial u}{\partial r} - i\lambda u\right) = 0 \tag{4.2.8}$$

where $f(\underset{\sim}{x}) = -\frac{\partial}{\partial \nu}(e^{i\lambda z} + v(\underset{\sim}{x}))$. We will now show how the functions $v(\underset{\sim}{x})$ and $u(\underset{\sim}{x})$ can be constructed by means of a "method of ascent". We make the assumption that $B(r)$ is a real valued continuously differentiable function of r for $r > 0$ with compact support contained in the interval $[0,a]$ where $a > 0$, and that D is bounded and strictly starlike with respect to the origin. In order to establish a "method of ascent" we consider the equation

$$\Delta_n u + \lambda^2(1+B(r))u = 0 \tag{4.2.9}$$

in place of (4.2.6) and later on set n=3.

We now look for a twice continuously differentiable solution $u(\underset{\sim}{x})$ of (4.2.9) defined in the exterior of D in the form

$$\begin{aligned} u(r,\theta) &= (\underset{\sim}{I}+\underset{\sim}{K})h \\ &= h(r,\theta) + \int_r^\infty s^{n-3}K(r,s;\lambda)h(s,\theta)ds \end{aligned} \tag{4.2.10}$$

where $(r,\theta) = (r,\theta_1,\ldots\theta_{n-1})$ are spherical coordinates, $h(r,\theta)$ is a twice continuously differentiable solution of

$$\Delta_n h + \lambda^2 h = 0 \tag{4.2.11}$$

in the exterior of D,and $K(r,s;\lambda)$ is a function to be determined. We assume

$$K(r,s;\lambda) = 0 \quad \text{for} \quad rs \geqslant a^2 \tag{4.2.12}$$

and note that if $h(r,\theta)$ satisfies the Sommerfeld radiation condition

$$\lim_{r\to\infty} r^{(n-1)/2}\left(\frac{\partial u}{\partial r} - i\lambda u\right) = 0 \quad , \tag{4.2.13}$$

then by (4.2.12) so will $u(r,\theta)$. We now substitute (4.2.10) into (4.2.9) and integrate by parts using (4.2.12). The result of this calculation is that (4.2.10) will be a solution of (4.2.9) provided $K(r,s;\lambda)$ is a twice continuously differentiable solution of

$$r^2\left[K_{rr} + \frac{n-1}{r}K_r + \lambda^2(1+B(r))K\right] = s^2\left[K_{ss} + \frac{n-1}{s}K_s + \lambda^2 K\right] \tag{4.2.14}$$

for $s > r$ satisfying (4.2.12) and the initial condition

$$K(r,r;\lambda) = -\frac{1}{2}\lambda^2 r^{2-n}\int_r^\infty sB(s)ds \quad . \tag{4.2.15}$$

Now let

$$\begin{aligned} \xi &= \log r \\ \eta &= \log s \end{aligned} \tag{4.2.16}$$

and define $M(\xi,\eta;\lambda)$ by

$$M(\xi,\eta;\lambda) = \exp\left[\left(\frac{n-2}{2}\right)(\xi+\eta)\right]K(e^\xi,e^\eta;\lambda), \tag{4.2.17}$$

i.e.

$$K(r,s;\lambda) = (rs)^{-(\frac{n-2}{2})} M(\log r, \log s;\lambda) . \tag{4.2.18}$$

Then $M(\xi,\eta;\lambda)$ satisfies the differential equation

$$M_{\xi\xi}-M_{\eta\eta} + \lambda^2(e^{2\xi}-e^{2\eta}+e^{2\xi}B(e^{\xi}))M = 0 \tag{4.2.19}$$

for $\eta > \xi$ and the auxilliary conditions

$$M(\xi,\xi;\lambda) = -\frac{1}{2}\lambda^2 \int_{\xi}^{\infty} e^{2\tau}B(e^{\tau})d\tau \tag{4.2.20}$$

$$M(\xi,\eta;\lambda) = 0 \quad \text{for } \frac{1}{2}(\xi+\eta) \geqslant \log a. \tag{4.2.21}$$

We assume that in addition to (4.2.19)-(4.2.21),

$$M(\xi,\eta;\lambda) = 0 \quad \text{for } \xi > \eta \quad . \tag{4.2.22}$$

Note that $M(\xi,\eta;\lambda)$, if it exists, is independent of the dimension n, and in this sense the operator (4.2.10) can be described as a "method of ascent".

We now proceed to construct a solution of (4.2.19)-(4.2.22). Our approach resemble that of section 2.1 for the operator $\underset{\sim}{A}_1$. Let

$$\begin{aligned} x &= \frac{1}{2}(\xi+\eta) \\ y &= \frac{1}{2}(\xi-\eta) \end{aligned} \tag{4.2.23}$$

and define $\tilde{M}(x,y;\lambda)$ by

$$\tilde{M}(x,y;\lambda) = M(x+y, x-y;\lambda) \quad . \tag{4.2.24}$$

The $\tilde{M}(x,y;\lambda)$ satisfies

$$\tilde{M}_{xy} - \lambda^2F(x+y,x-y)\tilde{M} = 0 \quad ; \quad y < 0 \tag{4.2.25}$$

$$\tilde{M}(x,0;\lambda) = -\frac{1}{2}\lambda^2 \int_{x}^{\infty} e^{2\tau}B(e^{\tau})d\tau \tag{4.2.26}$$

$$\tilde{M}(x,y;\lambda) = 0 \quad \text{for } x \geqslant \log a \tag{4.2.27}$$

$$\tilde{M}(x,y;\lambda) = 0 \quad \text{for } y > 0 \quad , \tag{4.2.28}$$

where in (4.2.25)

$$F(\xi,\eta) = -\left[e^{2\xi}-e^{2\eta}+e^{2\xi}B(e^{\xi})\right] \quad . \tag{4.2.29}$$

For $y \leqslant 0$, (4.2.25)-(4.2.27) imply that $\tilde{M}(x,y;\lambda)$ is the solution of the integral equation

$$\tilde{M}(x,y;\lambda) = -\frac{1}{2}\lambda^2\int_x^\infty e^{2\tau}B(e^\tau)d\tau \tag{4.2.30}$$
$$-\lambda^2\int_y^\infty\int_x^\infty F(\alpha+\beta,\alpha-\beta)\tilde{M}(\alpha,\beta;\lambda)d\alpha d\beta.$$

Note that (4.2.28) implies that the solution of the integral equation (4.2.30) satisfies the initial condition (4.2.26), and (4.2.27), (4.2.28), and the fact that B(r) has compact support guarnatee the existence of the integrals appearing in (4.2.30). Now in (4.2.30) make the change of variables

$$\alpha = \frac{1}{2}(\tau+\mu)$$
$$\beta = \frac{1}{2}(\tau-\mu) \quad . \tag{4.2.31}$$

Then (4.2.30) becomes

$$M(\xi,\eta;\lambda) = -\frac{1}{2}\lambda^2\int_{\frac{1}{2}(\xi+\eta)}^\infty e^{2\tau}B(e^\tau)d\tau \tag{4.2.32}$$
$$-\frac{1}{2}\lambda^2\int_\xi^\infty\int_{\eta+\xi-\tau}^{\eta+\tau-\xi} F(\tau,\mu)M(\tau,\mu;\lambda)d\mu d\tau.$$

Now note that in (4.2.32) if $\eta+\xi-\tau > \tau$, then $\mu > \tau$, and hence $M(\tau,\mu;\lambda)$ is not identically zero. On the other hand if $\eta+\xi-\tau < \tau$, then μ may be less than τ, and in such cases $M(\tau,\mu;\lambda) = 0$. Taking these facts into consideration we have that , for $\eta > \xi$, $M(\xi,\eta;\lambda)$ is the solution of the integral equation

$$M(\xi,\eta;\lambda) = -\frac{1}{2}\lambda^2\int_{\frac{1}{2}(\xi+\eta)}^\infty e^{2\tau}B(e^\tau)d\tau$$
$$-\frac{1}{2}\lambda^2\int_\xi^{\frac{1}{2}(\xi+\eta)}\int_{\eta+\xi-\tau}^{\eta+\tau-\xi} F(\tau,\mu)M(\tau,\mu;\lambda)d\mu d\tau \tag{4.2.33}$$
$$-\frac{1}{2}\lambda^2\int_{\frac{1}{2}(\xi+\eta)}^\infty\int_\tau^{\eta+\tau-\xi} F(\tau,\mu)M(\tau,\mu;\lambda)d\mu d\tau \quad .$$

We now want to solve (4.2.33) through the method of successive approximations. We look for a solution of (4.2.33) in the form

$$M(\xi,\eta;\lambda) = \sum_{j=0}^{\infty} M_j(\xi,\eta;\lambda) \tag{4.2.34}$$

where

$$M_o(\xi,\eta;\lambda) = -\frac{1}{2}\lambda^2 \int_{\frac{1}{2}(\xi+\eta)}^{\log a} e^{2\tau}B(e^{\tau})d\tau$$

$$\begin{aligned} M_j(\xi,\eta;\lambda) = &-\frac{1}{2}\lambda^2 \int_{\xi}^{\frac{1}{2}(\xi+\eta)} \int_{\eta+\xi-\tau}^{\eta+\tau-\xi} F(\tau,\mu)M_{j-1}(\tau,\mu;\lambda)d\mu d\tau \\ &-\frac{1}{2}\lambda^2 \int_{\frac{1}{2}(\xi+\eta)}^{\log a} \int_{\tau}^{\eta+\tau-\xi} F(\tau,\mu)M_{j-1}(\tau,\mu;\lambda)d\mu d\tau \end{aligned} \tag{4.2.35}$$

for $j \geqslant 1$. Note that the region of integration in (4.2.35) is only in the half-space $\frac{1}{2}(\xi+\eta) \leqslant \log a$ since $M_o(\xi,\eta;\lambda) = 0$ for $\frac{1}{2}(\xi+\eta) \geqslant \log a$ and this implies that for $\frac{1}{2}(\xi+\eta) \geqslant \log a$, $M_j(\xi,\eta;\lambda) = 0$ for each j. Assume $\eta \geqslant \xi \geqslant -\xi_o$ where ξ_o is a positive constant, and let

$$C = \frac{1}{2}|\lambda|^2 \max_{\substack{-\xi_o\leqslant\xi\leqslant\log a \\ -\xi_o\leqslant\eta\leqslant\xi_o+\log a}} \{\, e^{2\xi}|B(e^{\xi})|,\ |F(\xi,\eta)|\,\}. \tag{4.2.36}$$

Then for $\eta \geqslant \xi \geqslant -\xi_o$, $\frac{1}{2}(\xi+\eta) \leqslant \log a$, we have

$$\begin{aligned} |M_o(\xi,\eta;\lambda)| &\leqslant C(\log a - \tfrac{1}{2}(\xi+\eta)) \\ &\leqslant C(\log a - \xi) \end{aligned} \tag{4.2.37}$$

and

$$\begin{aligned} |M_1(\xi,\eta;\lambda)| \leqslant\ & 2C^2 \int_{\xi}^{\frac{1}{2}(\xi+\eta)} (\log a-\tau)(\tau-\xi)d\tau \\ &+C^2 \int_{\frac{1}{2}(\xi+\eta)}^{\log a} (\log a-\tau)(\tau-\xi)d\tau \quad . \end{aligned} \tag{4.2.38}$$

But in the second integral on the right hand side of (4.2.38) we have

$\frac{1}{2}(\xi+\eta) \leqslant \tau$, which implies $\eta \leqslant 2\tau - \xi$, and hence $\eta-\xi \leqslant 2(\tau-\xi)$. Therefore from (4.2.38) we have

$$|M_1(\xi,\eta;\lambda)| \leqslant 2C^2 \int_\xi^{\log a} (\log a-\tau)(\tau-\xi)d\tau . \tag{4.2.39}$$

But for $j \geqslant 0$ we have

$$\frac{1}{(2j+1)!} \int_\xi^{\log a} (\log a-\tau)^{2j+1}(\tau-\xi)d\tau = \frac{(\log a-\xi)^{2j+3}}{(2j+3)!} \tag{4.2.40}$$

and hence

$$|M_1(\xi,\eta;\lambda)| \leqslant \frac{2C^2}{3!} (\log a-\xi)^3. \tag{4.2.41}$$

By induction we have

$$|M_j(\xi,\eta;\lambda)| \leqslant \frac{2C^{j+1}}{(2j+1)!} (\log a-\xi)^{2j+1} \tag{4.2.42}$$

$$\leqslant \frac{2C^{j+1}}{(2j+1)!} (\log a+\xi_o)^{2j+1}$$

for $j \geqslant 0$, and hence the series (4.2.34) is absolutely and uniformly convergent for $\eta \geqslant \xi \geqslant -\xi_o$. This establishes the existence of the function $M(\xi,\eta;\lambda)$ and hence the kernel $K(r,s;\lambda)$. It is easily seen that since $B(r)$ is continuously differentiable, $K(r,s;\lambda)$ is twice continuously differentiable for $s \geqslant r > 0$. We note that $M(\xi,\eta;\lambda)$ is an entire function of λ and that

$$\lambda^{-2-2j}M_j(\xi,\eta;\lambda) = N_j(\xi,\eta) \tag{4.2.43}$$

is independent of λ. In particular $s^{n-3}K(r,s;\lambda)$ has the Taylor expansion

$$s^{n-3}K(r,s;\lambda) = s^{(n-4)/2}r^{(2-n)/2} \sum_{j=0}^{\infty} \lambda^{2j+2} N_j(\log r, \log s) \tag{4.2.44}$$

which is uniformly convergent for all complex values of λ. Note also that since $\underset{\sim}{K}$ is a Volterra operator, $(\underset{\sim}{I}+\underset{\sim}{K})^{-1}$ exists, in particular we can conclude (using (4.2.12), (4.2.14) and (4.2.15)) that for every solution

$u(\underset{\sim}{x})$ of (4.2.9) defined in the exterior of D there exists a solution $h(\underset{\sim}{x})$ of (4.2.11) defined in the exterior of D such that $u(\underset{\sim}{x})=(\underset{\sim}{I}+\underset{\sim}{K})h$. We summerize our results in the following theorem:

Theorem 4.2.1 ([20]): Let $u(\underset{\sim}{x})$ be a twice continuously differentiable solution of (4.2.9) in the exterior of D where D is strictly starlike with respect to the origin. Then $u(\underset{\sim}{x})$ can be represented in the form $u(\underset{\sim}{x})=(\underset{\sim}{I}+\underset{\sim}{K})h$ where $h(\underset{\sim}{x})$ is a twice continuously differentiable solution of (4.2.11) in the exterior of D. Conversely if $h(\underset{\sim}{x})$ is a solution of (4.2.11) in the exterior of D, then $u(\underset{\sim}{x})=(\underset{\sim}{I}+\underset{\sim}{K})h$ is a solution of (4.2.9) in the exterior of D. $u(\underset{\sim}{x})$ satisfies the Sommerfeld radiation condition (4.2.13) if and only if $h(\underset{\sim}{x})$ satisfies this condition.

We now want to use the integral operator $\underset{\sim}{I}+\underset{\sim}{K}$ (for n=3) to construct the functions $v(\underset{\sim}{x})$ and $u(\underset{\sim}{x})$ described in the introduction to this section. Since $e^{i\lambda z}$ is a solution of (4.2.11) we have that

$$w(\underset{\sim}{x}) = (\underset{\sim}{I}+\underset{\sim}{K})e^{i\lambda z} \tag{4.2.45}$$

is a solution of (4.2.9), and from (4.2.12) it is seen that we can choose $v(\underset{\sim}{x})$ to be

$$v(\underset{\sim}{x}) = \underset{\sim}{K}\, e^{i\lambda z} \quad . \tag{4.2.46}$$

To construct a solution $u(\underset{\sim}{x})$ of (4.2.6)-(4.2.8) we will use the operator $\underset{\sim}{I}+\underset{\sim}{K}$ in conjunction with the work of D.S.Jones on the exterior Neumann problem for the Helmholtz equation (4.2.11) (c.f. [38]). To describe the work of Jones, let $\lambda_1,\lambda_2,\ldots,\lambda_j,\ldots$ be the eigenvalues of the interior Dirichlet problem for (4.2.11) in D (for n=3). Then Jones has shown that if $\lambda < \lambda_{M+2}$ and $h(\underset{\sim}{x})$ is a solution of (4.2.11) (for n=3) satisfying prescribed Neumann data on ∂D and the Sommerfeld radiation condition (4.2.8) at infinity, there exists a continuous density $\psi(\underset{\sim}{x})$ such that $h(\underset{\sim}{x})$ can be represented in the form

$$h(\underset{\sim}{x}) = \int_{\partial D} \psi(\underset{\sim}{\xi})\Gamma(\underset{\sim}{\xi},\underset{\sim}{x};\lambda)\,d\omega_{\underset{\sim}{\xi}} \quad . \tag{4.2.47}$$

In (4.2.47)

$$\Gamma(\underset{\sim}{\xi},\underset{\sim}{x};\lambda) = \frac{e^{i\lambda R}}{R} + \sum_{m=0}^{M} \sum_{n=-m}^{m} b_{mn}\psi_{mn}(\underset{\sim}{x})\psi_{mn}(\underset{\sim}{\xi}) , \tag{4.2.48}$$

$R=|\underset{\sim}{x}-\underset{\sim}{\xi}|$, $d\omega_{\underset{\sim}{\xi}}$ is an element of surface area at the point $\underset{\sim}{\xi}\varepsilon\partial D$, the b_{mn} are nonzero real constants (arbitrary, but fixed), and

$$\psi_{mn}(\underset{\sim}{x}) = h_m^{(1)}(\lambda|\underset{\sim}{x}|)S_{mn}\left(\frac{\underset{\sim}{x}}{|\underset{\sim}{x}|}\right) \tag{4.2.49}$$

where $h_m^{(1)}$ denotes a spherical Hankel function and S_{mn} a spherical harmonic. Note that if the finite sum in (4.2.48) is not present, then in general it is not possible to represent $h(\underset{\sim}{x})$ in the form of the single layer potential (4.2.47) (c.f.[49]). Jones has also shown that for a given λ a suitable value of M can be chosen as follows: Let $\mu_1,\ldots,\mu_j,\ldots$ be the eigenvalues of the interior Dirichlet problem for (4.2.11) (for n=3) in the unit sphere (which can be computed from a knowledge of the zeros of the spherical Bessel functions - for a table of these zeros see [44]) and let r_o be the radius of the smallest sphere contained in D and r_1 the radius of the largest sphere containing D. Then

$$\frac{\mu_j}{r_o} \geqslant \lambda_j \geqslant \frac{\mu_j}{r_1} \quad . \tag{4.2.50}$$

In order to construct a solution $u(\underset{\sim}{x})$ of (4.2.6)-(4.2.8) we will look for a solution in the form

$$u(\underset{\sim}{x}) = (\underset{\sim}{I}+\underset{\sim}{K})h \tag{4.2.51}$$

where $h(\underset{\sim}{x})$ is a solution of (4.2.11) (for n=3) having the representation (4.2.47) in terms of an unknown continuous density $\psi(\underset{\sim}{x})$ to be determined. Note that $h(\underset{\sim}{x})$, and hence $u(\underset{\sim}{x})$, satisfies the Sommerfeld radiation condition (4.2.8). Substituting (4.2.47) into (4.2.51) and interchanging the orders of integration gives

$$u(\underset{\sim}{x}) = \int_{\partial D} \psi(\underset{\sim}{\xi})\Gamma(\underset{\sim}{\xi},\underset{\sim}{x};\lambda)d\omega_{\underset{\sim}{\xi}} \tag{4.2.51}$$

$$+ \int_{\partial D} \psi(\underset{\sim}{\xi}) \left\{ \int_{|\underset{\sim}{x}|}^{\infty} K(|\underset{\sim}{x}|,s;\lambda)\Gamma(\underset{\sim}{\xi}, s \frac{\underset{\sim}{x}}{|\underset{\sim}{x}|};\lambda)ds \right\} d\omega_{\underset{\sim}{\xi}} .$$

As in section 4.1 one can show that for $\underset{\sim}{x}$, $\underset{\sim}{\xi}$ on ∂D

$$\left| \frac{\partial}{\partial \nu_{\underset{\sim}{x}}} \left\{ \int_{|\underset{\sim}{x}|}^{\infty} K(|\underset{\sim}{x}|,s;\lambda)\Gamma(\underset{\sim}{\xi},s \frac{\underset{\sim}{x}}{|\underset{\sim}{x}|};\lambda)ds \right\} \right| \leq \frac{\text{constant}}{|\underset{\sim}{x}-\underset{\sim}{\xi}|} \tag{4.2.52}$$

where $\frac{\partial}{\partial \nu_{\underset{\sim}{x}}}$ denotes differentiation with respect to $\underset{\sim}{x}$ in the direction of the outward normal at $\underset{\sim}{x}$. Now let $\underset{\sim}{x}\varepsilon\partial D$, evaluate (4.2.51) at $\underset{\sim}{x}^1 \varepsilon \mathbb{R}^3\setminus\bar{D}$, and apply the operator $\nu_{\underset{\sim}{x}}.\nabla$ to both sides of (4.2.51). Letting $\underset{\sim}{x}^1$ tend to $\underset{\sim}{x}$, and using (4.2.52) and the discontinuity properties of the derivatives of single layer potentials (c.f. [21], [29]), we arrive at the following integral equation for $\psi(\underset{\sim}{x})$:

$$-\frac{1}{2\pi} f(\underset{\sim}{x}) = \psi(\underset{\sim}{x}) - \frac{1}{2\pi} \int_{\partial D} \psi(\underset{\sim}{\xi}) \frac{\partial}{\partial \nu_{\underset{\sim}{x}}} \Gamma(\underset{\sim}{\xi},\underset{\sim}{x};\lambda)d\omega_{\underset{\sim}{\xi}}$$

$$- \frac{1}{2\pi} \int_{\partial D} \psi(\underset{\sim}{\xi}) \frac{\partial}{\partial \nu_{\underset{\sim}{x}}} \left\{ \int_{|\underset{\sim}{x}|}^{\infty} K(|\underset{\sim}{x}|,s;\lambda)\Gamma(\underset{\sim}{\xi},s \frac{\underset{\sim}{x}}{|\underset{\sim}{x}|};\lambda)ds \right\} d\omega_{\underset{\sim}{\xi}} \tag{4.2.53}$$

$$= (\underset{\sim}{I}-\underset{\sim}{T}(\lambda))\psi .$$

A contstructive method for determining the desired function $u(\underset{\sim}{x})$ can now be obtained if we can show that the Fredholm integral equation (4.2.53) with weakly singular kernel can be uniquely solved for the unknown density $\psi(\underset{\sim}{x})$, i.e. that the operator $\underset{\sim}{I}-\underset{\sim}{T}(\lambda)$ is invertible. We will accomplish this by proving two theorems. The first theorem below proceeds along classical lines (c.f. [49]) except for the conclusion, where we make use of the operator $\underset{\sim}{I}+\underset{\sim}{K}$.

Theorem 4.2.2 ([20]): Let $\lambda > 0$ and let $u(\underset{\sim}{x}) \in C^2(\mathbb{R}^3\backslash\bar{D}) \cap C^1(\mathbb{R}^3\backslash D)$ be a solution of (4.2.6) in the exterior of D satisfying the Sommerfeld radiation condition (4.2.8) at infinity and the boundary condition $\frac{\partial u}{\partial \nu} = 0$ (or $u=0$) on ∂D. Then $u(\underset{\sim}{x}) \equiv 0$ for $\underset{\sim}{x} \in \mathbb{R}^3\backslash D$.

Proof: Let Ω be a ball of radius $r > a$ (recalling that $B(r)=0$ for $r \geq a$). Then from Green's formula we have

$$\iint_{\Omega\backslash D} (u\Delta\bar{u}-\bar{u}\Delta u)\,dV = \int_{\partial D} \left(\bar{u}\frac{\partial u}{\partial \nu} - u\frac{\partial \bar{u}}{\partial \nu}\right) d\omega - \int_{\partial\Omega} \left(\bar{u}\frac{\partial u}{\partial r} - u\frac{\partial \bar{u}}{\partial r}\right) d\omega \tag{4.2.54}$$

where dV denotes an element of volume and $d\omega$ an element of surface area. Since λ and $B(r)$ are real and $\frac{\partial u}{\partial \nu} = \frac{\partial \bar{u}}{\partial \nu} = 0$ (or $u=\bar{u}=0$) on ∂D, we have from (4.2.54) that

$$\int_{\partial D} \left(\bar{u}\frac{\partial u}{\partial r} - u\frac{\partial \bar{u}}{\partial r}\right) d\omega = 0. \tag{4.2.55}$$

But, for $r > a$, $u(\underset{\sim}{x})$ is a solution of $\Delta_3 h + \lambda^2 h = 0$ satisfying the Sommerfeld radiation condition (4.2.3), and hence for $r > a$

$$u(\underset{\sim}{x}) = \sum_{m=o}^{\infty} \sum_{n=-m}^{m} a_{mn} h_m^{(1)}(\lambda|\underset{\sim}{x}|) S_{mn}\left(\frac{\underset{\sim}{x}}{|\underset{\sim}{x}|}\right) \tag{4.2.56}$$

where the series converges absolutely and uniformly for $|\underset{\sim}{x}| \geq a+\varepsilon$, $\varepsilon > 0$ (c.f. [49]). By the orthogonality of the functions $S_{mn}\left(\frac{\underset{\sim}{x}}{|\underset{\sim}{x}|}\right)$ over the unit sphere and the formula

$$\overline{h_m^{(1)}(\lambda r)} \frac{d}{dr} h_m^{(1)}(\lambda r) - h_m^{(1)}(\lambda r) \overline{\frac{d}{dr} h_m^{(1)}(\lambda r)} = \frac{4i}{\pi\lambda^2 r^2} \tag{4.2.57}$$

we have from (4.2.55) and (4.2.56) that

$$\sum_{m=0}^{\infty} \sum_{n=-m}^{m} |a_{mn}|^2 = 0 , \tag{4.2.58}$$

which implies that $u(\underset{\sim}{x})=0$ for $r > a$. Let $u(\underset{\sim}{x}) = (\underset{\sim}{I}+\underset{\sim}{K})h$ for $\underset{\sim}{x} \in \mathbb{R}^3\backslash\bar{D}$. Then from (4.2.12) and the fact that $h(\underset{\sim}{x})$ can be determined from $u(\underset{\sim}{x})$ by inverting an integral equation of Volterra type (which implies that $h(\underset{\sim}{x})$ has the same smoothness properties that $u(\underset{\sim}{x})$ does), we can conclude that $h(\underset{\sim}{x})\varepsilon C^2(\mathbb{R}^3\backslash\bar{D})\cap C^1(\mathbb{R}^3\backslash D)$ and $h(\underset{\sim}{x})=0$ for $r > a$. Hence, since twice continuously differentiable solutions of $\Delta_3 h+\lambda^2 h=0$ are analytic functions of their independent variables (c.f. [21], [29]), we can conclude that $h(\underset{\sim}{x})=0$ for $\underset{\sim}{x}\varepsilon\mathbb{R}^3\backslash D$, and hence $u(\underset{\sim}{x})=(\underset{\sim}{I}+\underset{\sim}{K})h=0$ for $\underset{\sim}{x}\varepsilon\ \mathbb{R}^3\backslash D$.

We can now establish the following result on the invertibility of the Fredholm operator $\underset{\sim}{I}-\underset{\sim}{T}(\lambda)$:

<u>Theorem 4.2.3</u> ([20]): Let $\lambda > 0$ and define the operator $\underset{\sim}{T}_o(\lambda)$ by

$$\underset{\sim}{T}_o(\lambda)\psi = \frac{1}{2\pi}\int_{\partial D} \psi(\underset{\sim}{\xi})\ \frac{\partial}{\partial\nu_{\underset{\sim}{x}}}\ \Gamma(\underset{\sim}{\xi},\underset{\sim}{x};\lambda)\,d\omega_{\underset{\sim}{\xi}} \quad ; \quad \underset{\sim}{x}\varepsilon\partial D.$$

Then $(\underset{\sim}{I}-\underset{\sim}{T}(\lambda))^{-1}$ exists if and only if $(\underset{\sim}{I}-\underset{\sim}{T}_o(\lambda))^{-1}$ exists (where all mappings are understood to be in the space C^o, the space of continuous functions over ∂D with the maximum norm).

<u>Proof</u>: Since $\underset{\sim}{T}(\lambda)$ and $\underset{\sim}{T}_o(\lambda)$ are integral operators with weakly singular kernels, the Fredholm alternative is valid. Now let ψ be a solution of $(\underset{\sim}{I}-\underset{\sim}{T}(\lambda))\psi= 0$. Then the potential defined by (4.2.47) generates by (4.2.51) a solution of (4.2.6) in the exterior of D such that $u(\underset{\sim}{x})$ satisfies the Sommerfeld radiation condition, and, since $(\underset{\sim}{I}-\underset{\sim}{T}(\lambda))\psi=0$, we have $\frac{\partial u}{\partial\nu} = 0$ for $\underset{\sim}{x}\varepsilon\partial D$. From Theorem 4.2.2 we can now conclude that $u(\underset{\sim}{x})=0$ in the exterior of D. By inverting the Volterra equation (4.2.51) we can conclude that $h(\underset{\sim}{x})=0$ in the exterior of D and hence $(\underset{\sim}{I}-\underset{\sim}{T}_o(\lambda))\psi=0$ for $\underset{\sim}{x}\varepsilon\partial D$. If $(\underset{\sim}{I}-\underset{\sim}{T}_o(\lambda))^{-1}$ exists then we can conclude that $\psi(\underset{\sim}{x})=0$, and hence by the Fredholm alternative $(\underset{\sim}{I}-\underset{\sim}{T}(\lambda))^{-1}$ exists.

Conversely, if ψ is a solution of $(\underset{\sim}{I}-\underset{\sim}{T}_o(\lambda))\psi=0$, then $h(\underset{\sim}{x})$ as defined by (4.2.47) is zero for $\underset{\sim}{x}\varepsilon\ \mathbb{R}^3\backslash D$ and hence from (4.2.51) $u(\underset{\sim}{x})=0$ for $\underset{\sim}{x}\varepsilon\ \mathbb{R}^3\backslash D$. Then $\frac{\partial u}{\partial\nu} = 0$ for $\underset{\sim}{x}\varepsilon\partial D$ and $(\underset{\sim}{I}-\underset{\sim}{T}(\lambda))\psi=0$. Hence if $(\underset{\sim}{I}-\underset{\sim}{T}(\lambda))^{-1}$ exists we can conclude that $\psi(\underset{\sim}{x})=0$ and it follows from the Fredholm alternative that $(\underset{\sim}{I}-\underset{\sim}{T}_0(\lambda))^{-1}$ exists.

From the previously described work of Jones we now have the following Corollary:

Corollary 4.2.1: Let M be such that $\lambda < \lambda_{M+2}$ where λ_j denotes the jth eigenvalue for the interior Dirichlet problem for $\Delta_3 h+\lambda^2 h=0$ in D. Then $(\underset{\sim}{I}-\underset{\sim}{T}(\lambda))^{-1}$ exists.

Remark: A similar approach to that described above can be used to solve the Dirichlet, Neumann, and Robin problems for solutions of (4.2.9) defined in the exterior of D for all $n \geq 2$.

In the next section we will discuss an inverse problem associated with (4.2.9), i.e. the problem of determining the unknown function B(r) when the behaviour of $u(\underset{\sim}{x})$ at infinity is known, as well as the shape of the scattering body D and the boundary conditions on ∂D. It should be noted that other inverse problems can also be considered, for example that of determining the scattering body D given the function B(r), the behaviour of $u(\underset{\sim}{x})$ at infinity, and the boundary conditions on ∂D(c.f. [8]). Such inverse problems are in general improperly posed in the sense that the solution does not depend continuously on the behaviour of $u(\underset{\sim}{x})$ at infinity and a solution will not exist for arbitrarily prescribed "far field" data. We will not discuss the regularization of such problems, but instead consider only the case when the "far field" pattern is known exactly and is such that a solution is known to exist. For further discussion of inverse scattering problems for acoustic waves we refer the reader to

D. Colton, A reflection principle for solutions to the Helmholtz equation and an application to the inverse scattering problem, to appear in *Glasgow Math. J.*

and the references contained in this paper. A discussion of the use of integral operators in the investigation of certain inverse problems in scattering theory can also be found in [8] and [30].

4.3 The Inverse Scattering Problem.

The inverse problem we will consider in this section has its origins in the following problem connected with the scattering of acoustic waves in a nonhomogeneous medium (c.f. section 4.2). Let an incoming plane acoustic wave of frequency ω moving in the direction of the z axis be scattered off a "soft" sphere Ω of radius one which is surrounded by a pocket of rarefied or condensed air in which the local speed of sound is given by $c(r)$ where $r=|\underset{\sim}{x}|$ for $\underset{\sim}{x} \in \mathbb{R}^3$. Let $u_s(\underset{\sim}{x})e^{i\omega t}$ be the velocity potential of the scattered wave and let r,θ,ϕ be spherical coordinates in $\mathbb{R}^3$. Then from a knowledge of the far field pattern

$$f(\theta,\phi;\lambda) = \lim_{r\to\infty} re^{-i\lambda r}u_s(\underset{\sim}{x}) \tag{4.3.1}$$

for $\lambda = \frac{\omega}{c_o}$ (where $c(r) = c_o$ = constant for $r \geqslant a > 1$) contained in some finite interval $[\lambda_0,\lambda_1]$, we would like to determine the unknown function $c(r)$. Under the assumption that $|\nabla c(r)|$ is small compared with $\lambda c(r)$, we can formulate this problem mathematically as follows (c.f. [19], [20]): Let $B(r) = \left(\frac{c_o}{c(r)}\right)^2-1$ and set $u_s(\underset{\sim}{x}) = v(\underset{\sim}{x}) + u(\underset{\sim}{x})$ where $u(\underset{\sim}{x})$ satisfies

$$\Delta_3 u + \lambda^2(1+B(r))u = 0 \text{ in } \mathbb{R}^3\backslash\Omega \tag{4.3.2}$$

$$u(\underset{\sim}{x}) = -(e^{i\lambda z} + v(\underset{\sim}{x})) \text{ on } \partial\Omega \tag{4.3.3}$$

$$\lim_{r\to\infty} r\left(\frac{\partial u}{\partial r} - i\lambda u\right) = 0 \tag{4.3.4}$$

and $v(\underset{\sim}{x})$ is such that $e^{i\lambda z} + v(\underset{\sim}{x})$ is a solution of (4.3.2) in $\mathbb{R}^3\backslash\Omega$ where $v(\underset{\sim}{x}) = 0$ for $r \geqslant a$. (Note that $v(\underset{\sim}{x})$ is not uniquely defined). Then given

$$f(\theta,\phi;\lambda) = \lim_{r\to\infty} re^{-i\lambda r}u(\underset{\sim}{x}) \tag{4.3.5}$$

we want to determine the function $B(r)$. We will solve this problem by using the operator $\underset{\sim}{I}+\underset{\sim}{K}$ constructed in the previous section (c.f. [19]).

Let $J_{n+\frac{1}{2}}(\lambda r)$ and $H^{(1)}_{n+\frac{1}{2}}(\lambda r)$ denote respectively a Bessel function and Hankel function of the first kind, and define $j_{n+\frac{1}{2}}(r)$ and $h_{n+\frac{1}{2}}(r)$ by

$$\begin{aligned} j_{n+\frac{1}{2}}(r) &= (\underset{\sim}{I}+\underset{\sim}{K})((\lambda r)^{-\frac{1}{2}}J_{n+\frac{1}{2}}(\lambda r)) \\ h_{n+\frac{1}{2}}(r) &= (\underset{\sim}{I}+\underset{\sim}{K})((\lambda r)^{-\frac{1}{2}}H^{(1)}_{n+\frac{1}{2}}(\lambda r)) \quad . \end{aligned} \tag{4.3.6}$$

Then from the representation (c.f. [25], p.64)

$$e^{i\lambda z} = \sqrt{\frac{\pi}{2\lambda r}} \sum_{n=0}^{\infty} (2n+1)i^n J_{n+\frac{1}{2}}(\lambda r)P_n(\cos\theta) \tag{4.3.7}$$

where $P_n(\cos\theta)$ denotes Legendre's polynomial, it is easily verified using (4.2.46) that the solution of (4.3.2)-(4.3.4) is given by

$$u(\underset{\sim}{x}) = u(r,\theta) = -\sqrt{\frac{\pi}{2}} \sum_{n=0}^{\infty} \frac{(2n+1)i^n j_{n+\frac{1}{2}}(1)}{h_{n+\frac{1}{2}}(1)} h_{n+\frac{1}{2}}(r)P_n(\cos\theta) \ . \tag{4.3.8}$$

Note that from Theorem 4.2.2 we can conclude that $h_{n+\frac{1}{2}}(1) \neq 0$, and the convergence of the series (4.3.8) for $1 \leqslant r < \infty$, $0 \leqslant \theta \leqslant \pi$ follows from (4.3.6) and standard estimates for Bessel functions and Legendre polynomials for large values of n (c.f. [25] p.22-23 and p.205). From the fact that

$$h_{n+\frac{1}{2}}(r) = (\lambda r)^{-\frac{1}{2}}H^{(1)}_{n+\frac{1}{2}}(\lambda r); \ r \geqslant a \tag{4.3.9}$$

and the asymptotic estimate ([25] p.85)

$$H^{(1)}_{n+\frac{1}{2}}(\lambda r) = (-i)^{n+1}\sqrt{\frac{2}{\pi\lambda r}}\, e^{i\lambda r}\left[1+0\left(\frac{1}{\lambda r}\right)\right] \tag{4.3.10}$$

we can conclude that the far field pattern $f(\theta,\phi;\lambda) = f(\theta;\lambda)$ is given by (c.f. [8])

$$f(\theta;\lambda) = \sum_{n=0}^{\infty} \frac{i(2n+1)j_{n+\frac{1}{2}}(1)}{\lambda\, h_{n+\frac{1}{2}}(1)} P_n(\cos\theta) \quad . \tag{4.3.11}$$

Recall once again that although the far field pattern $f(\theta;\lambda)$ is assumed to be known, the functions $j_{n+\frac{1}{2}}(r)$ and $h_{n+\frac{1}{2}}(r)$ are unknown since $B(r)$ is as of yet unknown. However if we expand $f(\theta,\lambda)$ in a Legendre series

$$f(\theta;\lambda) = \sum_{n=0}^{\infty} \tilde{a}_n(\lambda)P_n(\cos\theta) , \tag{4.3.12}$$

then from (4.3.11) and (4.3.13) we have

$$\begin{aligned} \frac{j_{n+\frac{1}{2}}(1)}{h_{n+\frac{1}{2}}(1)} &= a_n(\lambda) \\ &= \lambda^{2n+1}(a_{n0} + a_{n1}\lambda^2 + \dots) \\ &\quad +\lambda^{4n+2}(c_{n0} + c_{n1}\lambda^2 + \dots) \end{aligned} \tag{4.3.13}$$

where

$$a_n(\lambda) = \frac{\lambda\tilde{a}_n(\lambda)}{i(2n+1)} \tag{4.3.14}$$

are known analytic functions of λ. The fact that $a_n(\lambda)$ has a zero of order $2n+1$ at the origin follows from (4.2.44), (4.3.6) and the series representations (c.f. [25] p.4)

$$(\lambda r)^{-\frac{1}{2}}J_{n+\frac{1}{2}}(\lambda r) = \sqrt{\frac{1}{2}} \sum_{m=0}^{\infty} (-1)^m \frac{(\lambda r/2)^{2m+n}}{m!\,\Gamma(m+n+\frac{3}{2})} \tag{4.3.15}$$

$$(\lambda r)^{-\frac{1}{2}}H^{(1)}_{n+\frac{1}{2}}(\lambda r) = \sqrt{\frac{1}{2}} \sum_{m=0}^{\infty} (-1)^m \left[\frac{(\lambda r/2)^{2m+n}}{m!\,\Gamma(m+n+\frac{3}{2})} + i\,\frac{(-1)^{n+1}(\lambda r/2)^{2m-n-1}}{m!\,\Gamma(m-n+\frac{1}{2})}\right].$$

Equating the coefficients of λ^{2n+1} and λ^{2n+3} respectively in (4.3.13) we have for $n \geqslant 0$ (using (4.2.44))

$$a_{n0} = i\,\frac{(-1)^n\Gamma(-n+\frac{1}{2})(\frac{1}{2})^{2n+1}}{\Gamma(n+\frac{3}{2})} \tag{4.3.16}$$

and, for $n > 0$,

$$- \frac{(\frac{1}{2})^{n+2}}{\Gamma(n+\frac{5}{2})} + \int_1^\infty N_0(\log 1, \log s) \frac{(\frac{1}{2})^n s^{n-\frac{1}{2}}}{\Gamma(n+\frac{3}{2})} ds$$

$$= a_{n0}\, i \frac{(-1)^n (\frac{1}{2})^{-n+1}}{\Gamma(-n+\frac{3}{2})} + a_{n1}\, i \frac{(-1)^{n+1}(\frac{1}{2})^{-n-1}}{\Gamma(-n+\frac{1}{2})} \tag{4.3.17}$$

$$+ a_{n0}\, i \int_1^\infty N_0(\log 1, \log s) \frac{(-1)^{n+1}(\frac{1}{2})^{-n-1} s^{-n-\frac{3}{2}}}{\Gamma(-n+\frac{1}{2})} ds \quad .$$

The equation corresponding to (4.3.17) for n=0 is exactly the same except that the term $c_{00}/\Gamma(\frac{3}{2})$ is added to the right hand side. Note that the coefficient a_{no} is independent of B(r). From (4.2.43) and (4.2.35) we have

$$\int_1^\infty N_0(\log 1, \log s)\, s^m ds = -\frac{1}{2}\int_1^{a^2}\int_{s^{\frac{1}{2}}}^{a} \xi B(\xi) s^m d\xi ds$$

$$= -\frac{1}{2}\int_1^{a}\int_1^{\xi^2} \xi B(\xi) s^m ds d\xi \tag{4.3.18}$$

$$= - \frac{1}{2(m+1)}\int_1^a \xi^{2m+3} B(\xi)\, d\xi + \frac{1}{2(m+1)}\int_1^a \xi B(\xi) d\xi,$$

and hence using (4.3.16) and (4.3.18) we can rewrite (4.3.17) as

$$\mu_n = \int_1^a B(s)\left[s^{2n+2} + s^{-2n} - 2s\right] ds \tag{4.3.19}$$

where for n > 0

$$\mu_n = -(2n+1)\left[\frac{-(2n+1)}{(2n+3)(1-2n)} + a_{n1}\, i \frac{(-1)^{n+1}(\frac{1}{2})^{-2n-1}\Gamma(n+\frac{3}{2})}{\Gamma(-n+\frac{1}{2})}\right]. \tag{4.3.20}$$

For n=0, μ_0 is the same as defined above except that the term c_{00} is subtracted from the right hand side. The μ_n are known from the far field pattern, and hence the problem of determining the function B(r) has now been reduced to solving the generalized moment problem (4.3.19), (4.3.20). Note that if we assume that B(r) is real valued, then from (4.3.16), (4.3.17) we have that a_{n1} is purely imaginary, and hence μ_n is real for each $n, n=0,1,2,\ldots$

We will now assume the existence of a continuously differentiable function $B(r)$ such that (4.3.19), (4.3.20) is valid, and address ourselves to the problems of uniqueness and approximation in $L^2[1,a]$. We restrict ourselves solely to the problem of uniqueness and approximation, since it is assumed a priori that the sequence μ_n (or a_{n1}) is a (generalized) moment sequence for some function $B(r)$ to be determined and hence the existence of $B(r)$ is not in question. The basic problems of uniqueness and approximation can be settled by appealing to the following theorem:

Theorem 4.3.1 ([19]): The functions

$$P_n(r) = r^{2n+2} + r^{-2n} - 2r \quad ,$$

$n=0,1,2,\ldots$, are complete in $L^2[1,a]$.

Proof: Let $f(r)$ be a continuous function on the interval $[1,a]$. Since the space of continuous functions on $[1,a]$ is dense in $L^2[1,a]$, to prove the theorem it suffices to show that if

$$\int_1^a f(s)P_n(s)ds = 0 \tag{4.3.21}$$

for $n=0,1,2,\ldots$, then $f(r)=0$ for $r\varepsilon[1,a]$. For $r\varepsilon\ [\frac{1}{a},1]$ define $f(r)$ by

$$f(r) = r^{-4}f(\tfrac{1}{r}) \quad ; \quad r\varepsilon\ [\tfrac{1}{a},1] \ . \tag{4.3.22}$$

Then

$$\int_1^a f(s)s^{-2n}ds = \int_{1/a}^1 f(s)s^{2n+2}ds \tag{4.3.23}$$

and hence from (4.3.21) we have

$$0 = \int_1^a f(s)\left[P_n(s) - P_{n+1}(s)\right]ds$$

$$= \int_1^a f(s)\left[s^{2n+2} + s^{-2n} - s^{2n+4} - s^{-2n-2}\right]ds$$

$$= \int_{1/a}^{a} f(s)\left[s^{2n+2} - s^{2n+4}\right]ds \tag{4.3.24}$$

$$= \frac{1}{2}\int_{1/a^2}^{a^2} f(s^{\frac{1}{2}})\left[s^{\frac{1}{2}} - s^{\frac{3}{2}}\right]s^n ds$$

for n=0,1,2,... . Since the set $\{r^n\}_{n=0}^{\infty}$ is complete in $L^2[\frac{1}{a},a]$, we have from (4.3.24) that

$$f(r^{1/2})(r^{1/2} - r^{3/2})=0 \tag{4.3.25}$$

for $r\varepsilon[\frac{1}{a},a^2]$, and hence f(r)=0 for $r\varepsilon[1,a]$.

The theorem is now proved.

The uniqueness of the function B(r) follows immediately from the above theorem. Furthermore the function B(r) can be approximated in $L^2[1,a]$ be orthonormalizing the set $\{P_n(r)\}_{n=0}^{\infty}$ over the interval $[1,a]$ to obtain the orthonormal set $\{\phi_n(r)\}_{n=0}^{\infty}$ and then approximating B(r) in $L^2[1,a]$ by the function

$$B_N(r) = \sum_{n=0}^{N} b_n\phi_n(r) \tag{4.3.26}$$

where

$$b_n = \int_1^a \phi_n(s)B(s)ds \quad . \tag{4.3.27}$$

The coefficients b_n can be found by using (4.3.19), (4.3.20). If it is assumed that $B(r)\varepsilon C^1[1,a]$, then it can be concluded that $B_N(r)$ approximates B(r) pointwise almost everywhere on $[1,a]$ (c.f.[43]).

Appendix

A Numerical Example .

Throughout these lectures we have given reference in the literature where numerical examples of the use of integral operator methods to solve boundary and initial-boundary value problems can be found (In particular see [1] ,[3] , [4],[5],[13],[24],[33],[34],[47]). The numerical procedure consists basically in the following steps (in the case of the solution of boundary or initial-boundary value problems by means of a complete family of solutions).

1) Approximation of the kernel of the integral operator. Since the kernel is given by either a recursion or iteration scheme, approximations can be obtained by truncating the iteration (or recursion) process after a finite number of steps. Error estimates can be obtained by either using the estimates used to show the series for the kernel converges, or by qualitatively observing that the contributions to the kernel become negligible after a finite number of iteration (or recursion)steps. The qualitative approach is reasonably safe since in practice the coefficients of the differential equation are polynomials and the kernel of the integral operator converges at a steady rate (in general geometrically).
2) Construction of a complete family of solutions. Since the approximation to the kernel of the integral operator is a polynomial (if the coefficients of the differential equation are polynomials) and so is the function operated on (i.e. z^n, $z^n t^m$ or the heat polynomial $h_n(x,t)$) this is merely a question of multiplying one polynomial by another and then performing the integration indicated in the definition of the integral operator.

3) Orthonormalization of the complete family by use of the Gram-Schmidt process, using numerical integration if necessary.

4) Computation of the Fourier coefficients. If the boundary of the domain is reasonably simple this (as well as step 3) can again be reduced to the "problem" of integrating the product of two polynomials (where we have assumed that the boundary or initial-boundary data has been approximated by polynomials).

5) Construction of the approximate solution and error estimates. Once one has completed step 4) the approximate solution of course follows immediately. Error estimates can be found by means of a priori estimates or (more simply) the maximum principle.
If one uses the integral operator in conjunction with double or single layer potentials to solve the desired boundary or initial-boundary value value problem then of course steps 2)-5) are replaced by

6) Solution of the integral equation. Since the resulting integral equations are of Volterra or Fredholm type, a numerical solution can be obtained by any one of a variety of known methods (c.f. [1], [22]).

7) Construction of the approximate solution. This is obtained by substituting the approximate density obtained from step 6) back into the double or single layer representation for the solution to the heat or Laplace or Helmholtz equation which the operator is operating on, constructing this solution to the heat or Laplace or Helmholtz equation, and then using the integral operator (and step 1)) to obtain the desired approximate solution to the original equation.

8) Error estimate. These are obtained from step 6), assuming the kernel of the integral operator has been approximated to a known degree of accuracy.
In the case of inverse problems, the numberical approach is similar in

spirit to the above steps for solving boundary value problems, but of course the details vary depending on the type of inverse problem being investigated.

The method for obtaining an analytic solution to various inverse problems has been given in these lectures (c.f. sections 1.5, 2.3, 3.3 and 4.3) and each of these is ameniable to numerical computations. In addition to approximating kernels of integral operators and the numerical integration of certain integrals, one must in some cases (e.g. in inverse problems in subsonic fluid flow and the inverse Stephan problem for the heat equation in two space variables) construct approximations to certain conformal mappings.

To illustrate the general approach for using integral operators to obtain numerical solutions to boundary or initial-boundary value problems we consider the following simple example due to Y.F. Chang of the Department of Computer Science, University of Nebraska (see also [13]). We want to use a complete family of solutions to construct an approximate solution to the initial-boundary value problem

$$u_{xx} - x^2 u = u_t \quad ; \quad -1 < x < 1,\ 0 < t < 1 \tag{1}$$

$$u(-1,t) = e^{-\frac{1}{2}-t}, \quad u(1,t) = e^{-\frac{1}{2}-t} \quad ; \quad 0 \leq t \leq 1$$
$$u(x,0) = e^{-\frac{1}{2}x^2} \; ; \; -1 \leq x \leq 1. \tag{2}$$

To construct a complete family of solutions we use the operator $\underset{\sim}{T}_3$ of section 2.1:

$$u(x,t) = \underset{\sim}{T}_3\{h\} = h(x,t) + \int_{-x}^{x} P(s,x)h(s,t)ds \tag{3}$$

where

$$P(s,x) = \frac{1}{2}\left[K(s,x) + M(s,x)\right] \tag{4}$$

is the (unique) solution of the initial value problem

$$P_{xx} - P_{ss} - x^2 P = 0 \tag{5}$$

$$P(x,x) = \frac{1}{2}\int_0^x s^2 ds = \frac{x^3}{6} \tag{6a}$$

$$P(-x,x) = 0 \tag{6b}$$

and $h(x,t)$ is a solution of

$$h_{xx} = h_t \quad . \tag{7}$$

Note that since the coefficients of (1) are independent of t, so is the kernel $P(s,x)$. The initial value problem (5),(6) satisfied by $P(s,x)$ follows from the initial value problems satisfied by $K(s,x)$ and $M(s,x)$ (c.f. (2.1.30), (2.1.31), (2.1.33), (2.1.34) and the facts that

$$\left.\begin{aligned} K(s,x) &= -K(-s,x) \\ M(s,x) &= M(-s,x). \end{aligned}\right) \tag{8}$$

From (5), (6) we have that $\tilde{P}(\xi,\eta) = P(\xi-\eta,\ \xi+\eta)$ can be constructed by the iterative scheme

$$\begin{aligned} \tilde{P}(\xi,\eta) &= \lim_{n\to\infty} \tilde{P}_n(\xi,\eta) \\ \tilde{P}_1(\xi,\eta) &= \frac{\xi^3}{6} \\ \tilde{P}_{n+1}(\xi,\eta) &= \frac{\xi^3}{6} + \int_0^\eta \int_0^\xi (\xi+\eta)^2 \tilde{P}_n(\xi,\eta)\, d\xi d\eta \end{aligned} \tag{9}$$

for $n=1,2,\ldots$. As an approximation to the kernel $P(s,x)$ we use $P_{10}(s,x)$ as defined by (9). A short calculation using (9) shows that

$$\max_{\substack{-1\leq x\leq 1 \\ 1\leq s\leq 1}} |P(s,x) - P_{10}(s,x)| \leq 1.6 \times 10^{-20}. \tag{10}$$

We now construct the (approximate) complete family of solutions

$$u_n(x,t) = h_n(x,t) + \int_{-x}^{x} P_{10}(s,x) h_n(s,t)\, ds \tag{11}$$

where

$$h_n(x,t) = n! \sum_{k=0}^{[\frac{n}{2}]} \frac{x^{n-2k} t^k}{(n-2k)!k!} \quad . \tag{12}$$

This is done for n=0,1,2, ... , 14. The integration in (11) is exact, i.e. the polynomials $P_{10}(s,x)$ and $h_n(s,t)$ are multiplied together, integrated, and added to $h_n(x,t)$. The set $\{u_n(x,t)\}_{n=0}^{14}$ is now orthonormalized over the base and vertical sides of the rectangle $-1 \leq x \leq 1$, $0 \leq t \leq 1$ by means of the Gram-Schmidt process to obtain the set $\{\phi_n(x,t)\}_{n=0}^{14}$. This is done using double precision arithmetic. The inner product used is

$$(\phi,\psi) = \int_0^1 \phi(-1,t)\psi(-1,t)dt + \int_{-1}^1 \phi(x,0)\psi(x,0)dx + \int_0^1 \phi(1,t)\psi(1,t)dt \quad . \tag{13}$$

The integrations performed in the Gram-Schmidt process are again exact, i.e. polynomials are multiplied together and integrated. The solution to the initial-boundary value problem (1),(2) is now approximated by the sum

$$u^*(x,t) = \sum_{n=0}^{14} a_n\phi_n(x,t) \tag{14}$$

where

$$a_n = (u,\phi_n) \quad . \tag{15}$$

Note that the coefficients a_n, n=0,1, ... , 14, can be computed solely from a knowledge of the functions $\phi_n(x,t)$, n=0,1, ..., 14, and the initial-boundary data (2). The coefficients a_n are computed by truncating the Taylor series for the functions $e^{-\frac{1}{2}t}$ and $e^{-\frac{1}{2}x^2}$ to the same order of accuracy as (10), and then computing (15) exactly by multiplying the appropriate polynomials together and integrating. Note that although the initial-boundary value problem (1), (2) is two dimensional, due to the fact we are approximating by means of a complete family of solutions only one dimensional integrals need be computed. This is an advantage of the present approach over other methods, where integrations must be performed over the entire rectangle $-1 \leq x \leq 1$, $0 \leq t \leq 1$. Such an advantage is of course particularly important in higher dimensional problems. We finally note that since the solution of (1), (2) is an even function of x, the odd coefficients $a_1, a_3, \ldots, a_{13}$ in (14), (15) all turn out to be identically zero.

The exact solution of (1), (2) is

$$u(x,t) = e^{-\frac{1}{2}x^2-t} \quad . \tag{16}$$

In Table I below we give the values of $u^*(x,t)$ at selected grid points and also the relative error defined by

$$\text{relative error} = \frac{u^*(x,t)-u(x,t)}{u(x,t)} \quad . \tag{17}$$

TABLE I /

TABLE I

x	1	Approximate solution	Relative error
0	0	1.00000	-6.9580×10^{-9}
0.2	0	0.98020	-3.3762×10^{-9}
0.4	0	0.92312	4.2696×10^{-9}
0.6	0	0.83527	8.0803×10^{-9}
0.8	0	0.72615	-4.3613×10^{-10}
1.0	0	0.60653	-2.2466×10^{-8}
0	0.2	0.81873	4.0571×10^{-10}
0.2	0.2	0.80252	9.3730×10^{-10}
0.4	0.2	0.75578	2.7202×10^{-9}
0.6	0.2	0.68386	6.1830×10^{-9}
0.8	0.2	0.59452	1.1356×10^{-8}
1.0	0.2	0.49659	1.5536×10^{-8}
0	0.4	0.67032	2.2209×10^{-9}
0.2	0.4	0.65705	1.7415×10^{-9}
0.4	0.4	0.61878	3.2910×10^{-11}
0.6	0.4	0.55990	-3.6697×10^{-9}
0.8	0.4	0.48675	-1.0332×10^{-8}
1.0	0.4	0.40657	-2.0325×10^{-8}
0.	0.6	0.54881	-1.1541×10^{-9}
0.2	0.6	0.53794	-8.6797×10^{-10}
0.4	0.6	0.50662	3.4421×10^{-10}
0.6	0.6	0.45841	3.6095×10^{-9}
0.8	0.6	0.39852	1.0898×10^{-8}
1.0	0.6	0.33287	2.4115×10^{-8}
0	0.8	0.44933	2.7676×10^{-9}
0.2	0.8	0.44043	2.5721×10^{-9}
0.4	0.8	0.41478	1.5339×10^{-9}
0.6	0.8	0.37531	-1.8415×10^{-9}
0.8	0.8	0.32628	-1.0323×10^{-8}
1.0	0.8	0.27253	-2.7649×10^{-8}
0	1.0	0.36788	-7.3333×10^{-11}
0.2	1.0	0.36059	4.9411×10^{-10}
0.4	1.0	0.33960	2.3005×10^{-9}
0.6	1.0	0.30728	4.1011×10^{-9}
0.8	1.0	0.26714	-5.4443×10^{-9}
1.0	1.0	0.22313	-8.4473×10^{-8}

Since $u(x,t)$ and $u^*(x,t)$ are even functions of x, values of the approximate solution and relative error are only given for $0 \leqslant x \leqslant 1$, $0 \leqslant t \leqslant 1$.

Note that since each $\phi_n(x,t)$ is a solution of (1), the maximum error (in absolute value) occurs on the base or vertical sides of the rectangle $-1 \leq x \leq 1$, $0 \leq t \leq 1$; in this case at the points $(x,t)=(\pm 1,1)$, where the relative error is 8.4473×10^{-8} in absolute value.

The computation time to construction $u^*(x,t)$ (i.e. to find the Taylor coefficients of $\phi_n(x,t)$, the coefficients a_n, and to evaluate $u^*(x,t)$ at selected grid points) using the CDC 6600 computer was approximately six seconds.

References

1. K.Atkinson, The numerical solution of Fredholm integral equations of the second kind with singular kernels, Numer.Math.19(1972),248-259.
2. S.Bergman, Integral Operators in the Theory of Linear Partial Differential Equations, Springer-Verlag, Berlin, 1969.
3. S.Bergman, Determination of subsonic flows around profiles, in Proceedings of the First United States National Congress of Applied Mechanics (1952),705-713.
4. S.Bergman and B.Epstein, Determination of a compressible fluid flow past an oval shaped obstacle, J.Math.Physics 26(1948),105-122.
5. S.Bergman and J.G.Herriot, Numerical solution of boundary value problems by the method of integral operators, Numer.Math.7(1965),42-65.
6. S.Bergman and M.Schiffer, Kernel Functions and Elliptic Differential Equations in Mathematical Physics, Academic Press,NewYork,1953.
7. R.C.Buck, Advanced Calculus, McGraw-Hill,New York,1956.
8. D.Colton, Partial Differential Equations in the Complex Domain, Pitman Press Lecture Note Series, London, 1976.
9. D.Colton, The noncharacteristic Cauchy problem for parabolic equations in one space variable, SIAM J.Math.Anal.5(1974),263-272.
10. D.Colton, Integral operators and reflection principles for parabolic equations in one space variable, J.Diff.Eqns.15(1974),551-559.
11. D.Colton, Generalized reflection principles for parabolic equations in one space variable, Duke Math.J.41(1974),547-553.
12. D.Colton, The approximation of solutions to initial-boundary value

problems for parabolic equations in one space variable, Quart.App.Math., to appear.

13. D.Colton, Complete families of solutions for parabolic equations with analytic coefficients, SIAM J.Math.Anal. to appear.

14. D.Colton, Integral operators for parabolic equations and their application, in Constructive and Computational Methods for Differential and Integral Equations, Springer-Verlag Lecture Note Series, Vol.430, Berlin,1974,95-111.

15. D.Colton, The solution of initial-boundary value problems for parabolic equations by the method of integral operators, to appear.

16. D.Colton, Bergman operators for parabolic equations in two space variables, Proc.Amer.Math.Soc.38(1973),119-126.

17. D.Colton, Runge's theorem for parabolic equations in two space variables, Proc.Royal Soc.Edin., 73A(1975),307-315.

18. D.Colton, The inverse Stefan problem for the heat equation in two space variables, Mathematika 21(1974),282-286.

19. D.Colton, The inverse scattering problem for acoustic waves in a spherically stratified medium, Proc.Edin.Math.Soc., to appear.

20. D.Colton and W.Wendland, Constructive methods for solving the exterior Neumann problem for the reduced wave equation in a spherically symmetric medium, Proc.Royal Soc.Edin., to appear.

21. R.Courant and D.Hilbert, Methods of Mathematical Physics,Vol.II, Interscience,New York,1961.

22. L.M.Delves and J.Walsh, Numerical Solution of Integral Equations, Clarendon Press,Oxford,1974.

23. M.Eichler, On the differential equation $u_{xx}+u_{yy}+N(x)u=0$, Trans.Amer. Math.Soc.65(1949).

24. S.Eisenstat, On the rate of convergence of the Bergman-Vekua method for the numerical solution of elliptic boundary value problems SIAM J.Numer.Anal.11(1974),654-680.

25. A.Erdélyi, et.al., Higher Transcendental Functions Vol.II, McGraw-Hill, New York,1953.

26. A.Friedman, Partial Differential Equations of Parabolic Type, Prentice-Hall,Englewood Cliffs, New Jersey,1964.

27. A.Friedman, Partial Differential Equations, Holt,Rinehart and Winston, New York,1969.

28. B.A.Fuks, Special Chapters in the Theory of Analytic Functions of Several Complex Variables, American Mathematical Society, Providence, Rhode Island,1965.

29. P.Garabedian, Partial Differential Equations, John Wiley, New York,1964.

30. R.P.Gilbert, Function Theoretic Methods in Partial Differential Equations, Academic Press,New York,1969.

31. R.P. Gilbert, Constructive Methods for Elliptic Equations, Springer-Verlag Lecture Note Series, Vol.365,Berlin,1974.

32. R.P.Gilbert, The construction of solutions for boundary value problems by function theoretic methods, SIAM J.Math.Anal.1(1970),96-114.

33. R.P.Gilbert, Integral operator methods for approximating solutions of Dirichlet problems, in Iterationsverfahren, Numerishe Mathematik, Approximationstheorie, International Series of Numerical Mathematics, Vol.15, Birkhauser Verlag, Basel,1970,129-146.

34. R.P.Gilbert and P.Linz, The numerical solution of some elliptic boundary value problems by integral operator methods, in Constructive and Computational Methods for Solving Differential and Integral Equations, Springer-Verlag Lecture Note Series,Vol.430,Berlin,1974,237-252.

35. G.Hellwig, Partial Differential Equations, Blaisdell,New York, 1964.

36. C.D.Hill, Parabolic equations in one space variable and the non-characteristic Cauchy problem, Comm.Pure App.Math.20(1967),619-633.

37. C.D.Hill, A method for the construction of reflection laws for a parabolic equation, Trans.Amer.Math.Soc.133(1968),357-372.

38. D.S.Jones, Integral equations for the exterior acoustic problem, Quart.J.Mech.App.Math.27(1974),129-142.

39. M.Krzyzanski, Partial Differential Equations of Second Order,Vol.I. Polish Scientific Publishers, Warsaw,1971.

40. B.Levitan, Generalized Translation Operators and Some of Their Applications, Israel Program for Scientific Translations, Jerusalem, 1964.

41. R.V.Mises and M.Schiffer, On Bergman's integration method in two-dimensional compressible flow, in Advances in Applied Mechanics,Vol.I, Academic Press,New York,1948,249-285.

42. N.I.Muskhelishvili, Singular Integral Equations, P.Noordhoff, Gronigen,1951.

43. I.P.Natanson, Constructive Function Theory,Vol.II, Frederick Ungar Publishing Co. New York,1965.

44. F.W.J.Olver, Royal Society Mathematical Tables, Vol.7, Bessel Functions (III), University Press,Cambridge,1960.

45. P.C.Rosenbloom and D.V.Widder, Expansions in terms of heat polynomials and associated functions, Trans.Amer.Math.Soc. 92(1959),220-266.

46. W.Rundell and M.Stecher, A method of ascent for parabolic and pseudoparabolic equations, SIAM J.Math.Anal.,to appear.

47. N.L.Schryer, Constructive approximation of solutions to linear elliptic boundary value problems, SIAM J.Numer.Anal.9(1972),546-572.

48. M.Stecher, The non-characteristic Cauchy problem for parabolic equations in three space variables, SIAM J.Math.Anal., to appear.

49. I.N.Vekua, New Methods for Solving Elliptic Equations, John Wiley, New York,1967.

50. J.L.Walsh, Interpolation and Approximation by Rational Functions in the Complex Domain, American Mathematical Society, Providence,1965.

8480-77-22-23-24
5-41